SpringerBriefs in Molecular Science

For further volumes:
http://www.springer.com/series/8898

Jun-Jie Zhu · Jing-Jing Li · Hai-Ping Huang
Fang-Fang Cheng

Quantum Dots for DNA Biosensing

Springer

Jun-Jie Zhu
Fang-Fang Cheng
School of Chemistry
  and Chemical Engineering
Nanjing University
Nanjing
People's Republic of China

Jing-Jing Li
School of Medical Imaging
Xuzhou Medical College
Xuzhou
People's Republic of China

Hai-Ping Huang
Jiangxi University of Science
  and Technology
Ganzhou
People's Republic of China

ISSN 2191-5407   ISSN 2191-5415   (electronic)
ISBN 978-3-642-44909-3   ISBN 978-3-642-44910-9   (eBook)
DOI 10.1007/978-3-642-44910-9
Springer Heidelberg New York Dordrecht London

Library of Congress Control Number: 2013953633

Printed on acid-free paper

Springer is part of Springer Science+Business Media (www.springer.com)

Preface

The goal of this brief is to give a summary of recent advances in "Quantum Dots for DNA Biosensing." Deoxyribonucleic acid (DNA), as the genetic information carrier, encodes all the genetic instructions used in the development and functioning of all known living organisms (bacteria, plants, yeast, and animals) and even many viruses. Due to its important function, the DNA-related detections are important for many applications in clinical diagnosis, homeland defense, and environment monitoring. Particularly, with the rapid development of nanotechnology, the sensitive, selective, and multiplexed analysis of gene sequences and quantification of target sequences have been realized. Among many of these analytical methods, quantum dots (QDs or Qdots) play a vital role in these applications. QDs have attracted great interests from researchers because of their excellent fluorescent, electrochemical and photoelectrochemical properties, which could be potentially and actually have been widely applied in various research areas ranging from in vitro biosensing to intracellular and in vivo imaging.

This brief focuses on special applications of QDs in DNA biosensing based on their fluorescent, electrochemical, photoelectrochemical, and electrochemiluminescent properties. Details of the preparation and functionalization of quantum dots as well as the fabrication of DNA biosensors have also been introduced here. We summarize how their properties can be used in DNA biosensor design with examples. Furthermore, we show some new emerging quantum dots such as silicon dots, carbon dots, and graphene dots as well as an important alternative to QDs, metal nanoclusters and their applications in DNA biosensing after introducing the limitations of traditional QDs. This brief is suitable to be used as a supplement for graduate-level courses in analytical chemistry, life science, biochemistry, biotechnology, biomedical engineering, etc. It may also help young scientists to get an overview of this topic.

Jun-Jie Zhu
Jing-Jing Li
Hai-Ping Huang
Fang-Fang Cheng

Contents

Chapter 1
Introduction

Abstract DNA biosensors have been widely studied because of their importance in clinical diagnosis, homeland defense as well as environment monitoring. Specifically, with the development of nanotechnology, such as the emergence of quantum dots (QDs), numerous QD-based DNA biosensors have been fabricated successfully. In this chapter, the overview of DNA biosensing and QDs is given. Meanwhile, the superior properties of QDs for the preparation of DNA biosensor are also listed, such as optical, electrochemiluminescence (ECL), electrochemical, and photoelectrochemical properties.

Keywords DNA biosensing • Quantum dot • Fluorescence • Electrochemilum inescence • Electrochemical property • Photoelectrochemical property

1.1 Overview of DNA Biosensing

DNA, or deoxyribonucleic acid, is the genetic information-bearing material in humans and almost all other living organisms. Along with RNA and proteins, DNA is one of the three major macromolecules essential for all known forms of life. Genetic information in DNA is stored as a code made up of four chemical bases: adenine (A), thymine (T), guanine (G), and cytosine (C). Human genome DNA consists of about 3 billion bases. More than 99 % of those bases are the same among all individuals. The sequence of these bases determines the information necessary for building and maintaining an organism.

Due to its important function, DNA serves as a type of biomarkers for many diseases including various cancers. Thus, the DNA-related detections are important for many applications in clinical diagnosis, homeland defense, and environment monitoring. There has been tremendous increase in such applications over the past few years, especially with the development of nanotechnology [1–3]. The analysis of gene sequences and the detection of gene amount play a fundamental

J.-J. Zhu et al., *Quantum Dots for DNA Biosensing*, SpringerBriefs in Molecular Science, DOI: 10.1007/978-3-642-44910-9_1, © The Author(s) 2013

role in offering the possibility of performing reliable diagnosis even before the appearance of any symptoms of a disease. Small variations in the genome affect our predisposition to diseases, such as cancer and congenital genetic diseases [4, 5]. In environmental and food areas, the detection of specific DNA sequences can be used for the monitoring of genetically modified organism (GMO) or pathogenic bacteria [6].

Biosensors refer to self-containing integrated devices, being capable of providing specific quantitative or semiquantitative analytical information using a biological recognition element which is followed by contact with a transduction element. Major processes involved in a biosensor are specific target recognition event and signal transduction [7]. The first successfully developed DNA biosensor achieved rapid screening of toxic substances. The biosensor was constructed by immobilizing a double helix DNA (Calf Thymus DNA) onto screen-printed electrodes and then used for the determination of the toxicity of different kinds of common surfactants based on the height of the guanine oxidation peak. The interactions with toxic substances raise structural and conformational modifications of DNA causing decrease in guanine peak [8]. Except such kind of direct oxidation of guanine, the DNA probe must be thus chemically or enzymatically labeled with radioactive material, chemiluminophore, fluorophore, electrochemical, or photoelectrochemical materials to generate signal because the nucleic acid itself are not able to provide any signal. Quantum dots (QDs) emerge as excellent fluorescent nanomaterials possessing great potentials in DNA biosensor fabrication. Up to now, different types of DNA biosensors based on amperometric [9, 10], potentiometric [11, 12], piezoelectric [13, 14], thermal [15, 16], and optical techniques [17, 18] have been developed. In this book, we will focus on those DNA biosensors fabricated by taking advantages of properties of QDs (optical, electrochemical luminescence, electrochemical, and photoelectrochemical). Firstly, we overview different types of QDs. Preparation methods and functionalization of QDs were presented in Chap. 2. This provides a basis for the fabrication of QD-based DNA biosensors. Then, based on the properties of QDs (optical, electrochemiluminescence, electrochemical, and photoelectrochemical), we numerate these QD-based DNA and RNA biosensors in Chaps. 3, 4, and 5.

1.2 Overview of Quantum Dots

Quantum dots (QDs or Qdots) are colloidal semiconductor nanocrystals and have been developed for more than 30 years. Since discovered in glass crystals by Russian physicist Alexei Ekimov in 1980 [19], QDs have attracted great interests from researchers, demonstrated by the large number of scientific publications in this area (Fig. 1.1). QDs' excellent fluorescent, electrochemical, and photoelectrochemical properties have been widely applied in various research areas, ranging from in vitro biosensing to intracellular and in vivo imaging.

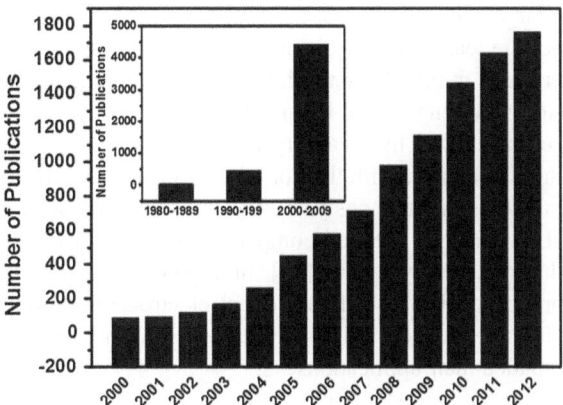

Fig. 1.1 Annual trends in the number of publications for QDs. The terms "Quantum Dots" and "QDs" have been considered. The literature search was done using PubMed

1.2.1 Optical Property

The size-tunable absorption and luminescence spectra of QDs arise from the quantum confinement effect. Excitons of QDs are confined in all three spatial dimensions and divided into discrete energy levels, which are similar to the energy levels in atoms. Electrons are first excited from the occupied level to the unoccupied energy level and fluorescence occurs when the excited electron relaxes to the ground state and combines with the hole. In a simplified model system, the energy of the emitted photon can be deemed as the sum of the bandgap energy between the occupied level and the unoccupied energy level, the confinement energies of the hole and the excited electron, and the bound energy of the exciton (the electron–hole pair). A decrease in the size of the QDs will bring an increase in the bandgap energy which represents a hypsochromic shift of the absorption and photoluminescence (PL) spectra [20, 21].

Since the emergence of QDs, comparison of pros and cons of QDs and organic fluorophores have always been a hot topic. Organic fluorophores have a narrow absorption spectrum which results in a narrow range of emission and they do not have a sharp symmetric emission peak which is further broadened by a red tail. In contrast, QDs have a broader excitation spectra and a more sharply defined emission peak. QDs are several thousand times more stable against photobleaching than organic dyes and are thus well suited for continuous tracking studies over a long period of time. One order of magnitude longer excited-state lifetime of QDs than that of organic dyes provides a feasible approach to separate the QD fluorescence from background fluorescence to achieve an accurate signal readout [22, 23]. Meanwhile, the large Stokes shifts of QDs (measured by the distance between the excitation and emission peaks) can be used to further improve detection sensitivity. As reported, the Stokes shifts of QDs can be as large as 300–400 nm, depending on the wavelength of the excitation light. Organic dye signals with a small Stokes shift are often buried by strong tissue autofluorescence, whereas QDs signals with a large Stokes shift can be clearly recognized over the background [24]. Last but not the least, another distinct property of QDs is that QDs with different sizes can produce multicolor emissions at the same excitation wavelength and their emission wavelengths can be tuned continuously by varying particle size and chemical composition. This is a very important feature for multiplexed detection or tracking. Especially, with the complexity of disease, tracking a panel of molecular biomarkers simultaneously is very important for accurate diagnosis [25–28]. Nie et al. provided a detailed comparison of such properties between QDs and organic fluorophores by figures [29]. In a word, QDs possess not only outstanding optical properties, but also the excellent electrochemiluminescence, electrochemical, and photoelectrochemical properties which further stimulate their fast development and broaden their applications.

1.2.2 Electrochemiluminescence Property

Electrochemiluminescence or electrogenerated chemiluminescence is a kind of luminescence produced during electrochemical reactions in solution. The first QDs ECL behavior was found by Bard et al. in 2002. Silicon QDs could generate light emission during potential cycling or pulsing [30]. Later, people found the elemental and compound semiconductors, such as Ge [31], CdTe [32], PbS [33], CdSe [34, 35], and ZnS [36], can also generate efficient ECL. The ECL mechanism of semiconductor QDs mainly depends on the annihilation or coreactant ECL reaction. Additionally, it is believed that the ECL emission is not sensitive to NP size and capping agent used but depends more sensitively on surface chemistry and the presence of surface states. The exact ECL mechanism of various QDs and their applications for DNA analysis can be found in Chap. 4.

1.2.3 Electrochemical and Photoelectrochemical Property

Although to a lesser extent compared to their optical properties, the electrochemical property of QDs has also been explored and applied in inorganic substance analysis, organics analysis, immunoassay, aptasensing assay, and solar cell. The electrochemical behavior of QDs revealed the quantized electronic behavior as well as decomposition reactions upon reduction and oxidation. Their electrochemical property is influenced by various factors, such as the QDs size, the capping stabilizer, the value of pH, and the coexisted chemicals [37]. Photoelectrochemically active species usually used are rutheniumbipyridine derivatives [38–41], semiconductor nanoparticles [42–45], and dyes [46, 47]. Photoexcitation of the semiconductor QDs results in the transfer of electrons from the valence band to the conduction band, thus yielding electron–hole pairs [48–50]. Whereas the luminescence properties of QDs originate from radiative

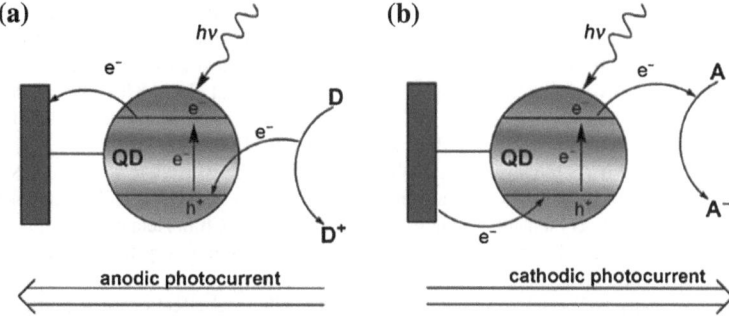

Fig. 1.2 Photocurrents generated by semiconductor NPs associated with electrodes. **a** Anodic photocurrent. **b** Cathodic photocurrent. Reproduced with permission from Ref. [51]. Copyright 2008, Wiley

electron–hole recombination, trapping of conduction-band electrons in surface traps yields sufficiently long-lived electron–hole pairs to permit the ejection of the trapped electrons to electrodes (or a solution-solubilized electron acceptor A) giving rise to the photoelectrochemical current. The ejection of the conduction-band electrons to the electrode, with the concomitant transfer of electrons from a solution-solubilized electron donor D, yields an anodic photocurrent (Fig. 1.2a). In contrast, transfer of the conduction-band electrons to a solution-solubilized electron acceptor, followed by the supply of electrons from the electrode to neutralize the valence-band holes, yields a cathodic photocurrent (Fig. 1.2b) [51]. QDs, with their unique fluorescence properties and photoelectrochemical functions, are photoactive materials for the development of nucleic acid sensor systems.

References

1. Ngo HT, Wang H-N, Fales AM, Vo-Dinh T (2013) Label-free DNA biosensor based on SERS molecular sentinel on nanowave chip. Anal Chem 85(13):6378–6383
2. Sun H, Choy TS, Zhu DR, Yam WC, Fung YS (2009) Nano-silver-modified PQC/DNA biosensor for detection *E. coli* in environmental water. Biosens Bioelectron 24(5):1405–1410
3. Anjum V, Pundir CS (2007) Biosensors: future analytical tools. Sens Transducers 76:937–944
4. Kruglyak L (1999) Prospects for whole-genome linkage disequilibrium mapping of common disease. Nat Genet 22:139–144
5. Knudson AG (2002) Cancer genetics. Am J Med Genet 111:96–102
6. Suman AK (2008) Recent advances in DNA biosensor. Sens Transducers J 92(5):122–133
7. Wang J (2000) From DNA biosensors to gene chips. Nucleic Acids Res 28(16):3011–3016
8. Francesca C (2011) DNA based biosensors for environmental and medical applications. University of Cagliari, Cagliari Doctoral Thesis
9. Babkina SS, Ulakhovich NA (2005) Complexing of heavy metals with DNA and new bio-affinity method of their determination based on amperometric DNA based biosensor. Anal Chem 77(17):5678–5685

10. Ho KC, Cheu CY, Hsu HC, Cheu LC, Shiesh SC, Lin XZ (2004) Amperometric detection of morphine at a prussian blue modified indium tin oxide electrode. Biosens Bioelectron 20:3–8
11. Ding J, Chen Y, Wang X, Qin W (2012) Label-free and substrate-free potentiometric apta-sensing using polycation-sensitive membrane electrodes. Anal Chem 84(4):2055–2061
12. Du M, Yang T, Jiao K (2010) Rapid DNA electrochemical biosensing platform for label-free potentiometric detection of DNA hybridization. Talanta 81(3):1022–1027
13. Fei Y, Jin XY, Wu ZS, Zhang SB, Shen G, Yu RQ (2011) Sensitive and selective DNA detection based on the combination of hairpin-type probe with endonuclease/GNP signal amplification using quartz-crystal-microbalance transduction. Anal Chim Acta 691(1–2):95–102
14. Bunde RL, Jarvi EJ, Rosentreter JJ (1998) Piezoelectric quartz crystal biosensors. Talanta 46(6):1223–1236
15. Tan Z-Q, Zhang N-H (2013) An analytical model for thermal effect of microcantilever-DNA biosensors. Int J Thermophys 34(6):1049–1065
16. Mao YD, Luo CX, Ouyang Q (2003) Studies of temperature-dependent electronic transduction on DNA hairpin loop sensor. Nucleic Acids Res 31(18):e108–e122
17. Zhang H, Jia Z, Lv X, Zhou J, Chen L, Liu R, Ma J (2013) Porous silicon optical microcavity biosensor on silicon-on-insulator wafer for sensitive DNA detection. Biosens Bioelectron 44:89–94
18. Valentini P, Fiammengo R, Sabella S, Gariboldi M, Maiorano G, Cingolani R, Pompa PP (2013) Gold-nanoparticle-based colorimetric discrimination of cancer-related point mutations with picomolar sensitivity. ACS Nano 7(6):5530–5538
19. Ekimov AI, Onushchenko AA (1981) Quantum size effect in three-dimensional micro-scopic semiconductor crystals. JETP Lett 34:345–349
20. Wang Y, Herron N (1991) Nanometer-sized semiconductor clusters: materials synthesis, quantum size effects, and photophysical properties. J Phys Chem 95:525–532
21. Alivisatos AP (1996) Semiconductor clusters, nanocrystals, and quantum dots. Science 271:933–937
22. Jakobs S, Subramaniam V, Schonle A, Jovin TM, Hell SW (2000) EGFP and DsRed expression cultures of *Escherichia coli* imaged by confocal, two-photon and fluorescence lifetime microscopy. FEBS Lett 479:131–135
23. Pepperkok R, Squire A, Geley S, Bastiaens PIH (1999) Simultaneous detection of multiple green fluorescent proteins in live cells by fluorescence lifetime imaging microscopy. Curr Biol 9:269–272
24. Gao XH, Cui YY, Levenson RM, Chung LWK, Nie SM (2004) In vivo cancer targeting and imaging with semiconductor quantum dots. Nat Biotechnol 22:969–976
25. Gao XH, Nie SM (2003) Molecular profiling of single cells and tissue specimens with quantum dots. Trends Biotechnol 21:371–373
26. Sidransky D (1997) Nucleic acid-based methods for the detection of cancer. Science 278:1054–1058
27. Jung R, Peterson K, Kruger W, Wolf M, Wagener C, Zander A (1999) Detection of micrometastasis by cytokeratin 20 RT-PCR is limited due to stable background transcription in granulocytes. Br J Cancer 81:870–873
28. Li N, Chang CY, Pan W, Tang B (2012) A multicolor nanoprobe for detection and imaging of tumor-related mRNAs in living cells. Angew Chem Int Ed 51:7426–7430
29. Gao XH, Yang L, Petros JA, Marshall FF, Simons JW, Nie SM (2005) In vivo molecular and cellular imaging with quantum dots. Curr Opin Biotechnol 16:63–72
30. Ding Z, Quinn BM, Haram SK, Pell LE, Korgel BA, Bard AJ (2002) Electrochemistry and electrogenerated chemiluminescence from silicon nanocrystal quantum dots. Science 296:1293–1297
31. Myung N, Lu X, Johnston KP, Bard AJ (2004) Electrogenerated chemiluminescence of Ge nanocrystals. Nano Lett 4:183–185
32. Bae Y, Myung N, Bard AJ (2004) Electrochemistry and electrogenerated chemiluminescence of CdTe nanoparticles. Nano Lett 4:1153–1161

33. Sun LF, Bao L, Hyun BR, Bartnik AC, Zhong YW, Reed JC, Pang DW, Abruña HD, Malliaras GG, Wise FW (2009) Electrogenerated chemiluminescence from PbS quantum dots. Nano Lett 9:789–793
34. Myung N, Ding Z, Bard AJ (2002) Electrogenerated chemiluminescence of CdSe nanocrystals. Nano Lett 2:1315–1319
35. Myung N, Bae Y, Bard AJ (2003) Effect of surface passivation on the electrogenerated chemiluminescence of CdSe/ZnSe nanocrystals. Nano Lett 3:1053–1055
36. Shen L, Cui X, Qi H, Zhang C (2007) Electrogenerated chemiluminescence of ZnS nanoparticles in alkaline aqueous solution. J Phys Chem C 111:8172–8175
37. Huang HP, Zhu J-J (2013) The electrochemical applications of quantum dots. Analyst 138:5855–5865
38. Liang MM, Liu SL, Wei MY, Guo L-H (2006) Photoelectrochemical oxidation of DNA by ruthenium tris (bipyridine) on a tin oxide nanoparticle electrode. Anal Chem 78:621–623
39. Liang MM, Guo L-H (2007) Photoelectrochemical DNA sensor for the rapid detection of DNA damage induced by styrene oxide and the fenton reaction. Environ Sci Technol 41:658–664
40. Liang MM, Jia SP, Zhu SC, Guo LH (2008) Photoelectrochemical sensor for the rapid detection of in situ DNA damage induced by enzyme-catalyzed fenton reaction. Environ Sci Technol 42:635–639
41. Wang L-R, Qu N, Guo L-H (2008) Electrochemical displacement method for the investigation of the binding interaction of polycyclic organic compounds with DNA. Anal Chem 80:3910–3914
42. Willner I, Patolsky F, Wasserman J (2001) Photoelectrochemistry with controlled DNA-cross-linked CdS nanoparticle arrays. Angew Chem Int Ed 40:1861–1864
43. Sheeney-Haj-Ichia L, Basnar B, Willner I (2005) Efficient generation of photocurrents by using CdS/carbon nanotube assemblies on electrodes. Angew Chem Int Ed 44:78–83
44. Hojeij M, Su B, Tan S, Mériguet G, Girault HH (2008) Nanoporous photocathode and photoanode made by multilayer assembly of quantum dots. ACS Nano 2:984–992
45. Li Y-J, Ma M-J, Yin G, Kong Y, Zhu J-J (2013) Phthalocyanine-sensitized graphene-CdS nanocomposites: an enhanced photoelectrochemical immunosensing platform. Chem Eur J 19:4496–4505
46. Cooper JA, Wu M, Compton RG (1998) Photoelectrochemical analysis of ascorbic acid. Anal Chem 70:2922–2927
47. Okamoto A, Kamei T, Tanaka K, Saito I (2004) Photostimulated hole transport through a DNA duplex immobilized on a gold electrode. J Am Chem Soc 126:14732–14733
48. Zhang XR, Xu YP, Yang YQ, Jin X, Ye SJ, Zhang SS, Jiang LL (2012) A new signal-on photoelectrochemical biosensor based on a graphene/quantum-dot nanocomposite amplified by the dual-quenched effect of bipyridinium relay and AuNPs. Chem Eur J 18:16411–16418
49. Bas D, Boyaci IH (2009) Quantitative photoelectrochemical detection of biotin conjugated CdSe/ZnS quantum dots on the avidin immobilized ITO electrodes. Electroanalysis 21(16):1829–1834
50. Zhao XM, Zhou SW, Jiang L-P, Hou WH, Shen QM, Zhu J-J (2012) Graphene-CdS nanocomposites: facile one-step synthesis and enhanced photoelectrochemical cytosensing. Chem Eur J 18:4974–4981
51. Gill R, Zayats M, Willner I (2008) Semiconductor quantum dots for bioanalysis. Angew Chem Int Ed 47:7602–7625

Chapter 2
Quantum Dots

Abstract In the past 30 years, quantum dots (QDs) have developed a lot from their kinds to the various application areas. Traditional nanocrystals are usually composed of elements from groups III–V, II–VI, or IV–VI of the periodic table, such as CdS, CdSe, CdTe, CdS@ZnS, CdSe@ZnS, CdSeTe@ZnS. These QDs own excellent fluorescence properties and have been widely used in biosensing and intracellular or in vivo imaging. However, the leaked cadmium ions are culprits for the observed cytotoxicity of cadmium-based QDs, which hampers their further practical applications. Later, with the demand for more biocompatible QDs as the signal reporter, cadmium-free quantum dots (CFQDs) were introduced, such as silicon QDs (Si QDs), carbon dots (C-dots), and graphene QDs (GQDs). In this chapter, the kinds of these traditional quantum dots and new emerging quantum dots as well as their preparation and functionalization are discussed in detail. Additionally, as a viable alternative to QDs, the metal nanoclusters also displayed great potentials as luminescent labels for fluorescent biosensing and bioimaging. Thus, the relevant description of metal nanoclusters is also included in this chapter.

Keywords Quantum dots • Metal nanoclusters • Preparation and functionalization • Bioconjugation

2.1 Traditional Quantum Dots

History of QDs begins with their first discovery in glass crystals in 1980 by Russian physicist Ekimov [1]. Systematic advancement in the science and technology of QDs was driven after 1984, when Luis Brus derived a relation between size and bandgap for semiconductor nanoparticles by applying a particle in a sphere model approximation to the wave function for bulk semiconductors [2, 3]. However, it took nearly a decade for a new promotion in QD research until the

J.-J. Zhu et al., *Quantum Dots for DNA Biosensing*, SpringerBriefs in Molecular Science, DOI: 10.1007/978-3-642-44910-9_2, © The Author(s) 2013

successful synthesis of colloidal CdX (X = S, Se, Te) QDs with size-tunable band-edge absorption and emissions by Murray et al. [4]. So far, CdX is the most investigated QDs due to their excellent optical and electrochemical properties. However, with the further application in biological area, the toxicity of cadmium ion in CdX was paid more and more attention. In order to improve the biocompatibility as well as the PL quantum yield and stability of these core nanocrystals, a layer of a few atoms with a higher bandgap semiconductor was introduced to encapsulate the core nanocrystals to form core–shell nanocrystals. The luminescence efficiency is significantly improved when the nanocrystals are passivated on their surface by a shell of a larger bandgap semiconductor and the leaching of metal ions from the core is blocked well by this structure [5, 6]. At the beginning, CdSe/ZnS and CdSe/CdS are the most intensively studied [5, 7]. Later, more and more other "core–shell" QDs were developed, such as CdSe/ZnSe [8], CdTe/CdS [9], CdTe/ZnS [10], and even CdTe/CdS/ZnS "core/shell/shell" QDs [11]. Reiss et al. proposed a simple synthetic route for the preparation of CdSe/ZnSe core/shell nanocrystals applying zinc stearate as a zinc source. Based on the literature, they firstly synthesized CdSe core nanocrystals in a mixed TOPO/HAD solvent with a molar ratio of 60–80 % HAD and using CdO, complexed by dodecylphosphonic acid, as cadmium precursor. Then, ZnO was complexed with dodecylphosphonic acid and slowly injected together with TOPSe into a mixture of HAD-/TOPO-containing CdSe core nanocrystals. After the formation of CdSe/ZnSe QDs, mercaptocarboxylic acids were introduced to make them water soluble and the photoluminescence efficiencies in organic solvents as well as in water after functionalization with mercaptoundecanoic acid could reach 60–85 % [8]. In our group, CdSeTe@ZnS-SiO$_2$ QDs were prepared with ZnS-like clusters filled into the SiO$_2$ shell via a microwave-assisted approach (shown in Fig. 2.1). The mercaptopropionic acid (MPA)-capped green-emitting CdSeTe alloy quantum dots were firstly prepared and purified. After that, the CdSeTe QDs were coated with a silica layer at room temperature in the presence of Zn^{2+} and glutathione (GSH). Lastly, the silica-coated QDs were refluxed under microwave irradiation. With the increase in reaction time, the fluorescence gradually changed from dim green to bright orange under a 365-nm excitation and quantum yield was enhanced from 11.9 to 56.9 % with rhodamine 6G as standard before and after the ZnS–SiO$_2$ coating [12].

2.2 New Emerging Quantum Dots

For traditional QDs, cadmium is the main element for their composition. However, it is well known that leaked cadmium ions are culprits for the observed cytotoxicity of cadmium-based QDs, which hampers their further applications to cellular or in vivo study. With the demand for more biocompatible QDs as the signal reporter, the emphasis has shift toward the fabrication of cadmium-free quantum dots (CFQDs) for applications in biology, such as silicon QDs (Si QDs), carbon dots (C-dots), graphene QDs (GQDs), Ag$_2$Se, Ag$_2$S, InP, CuInS$_2$/ZnS. Strictly, some of them are

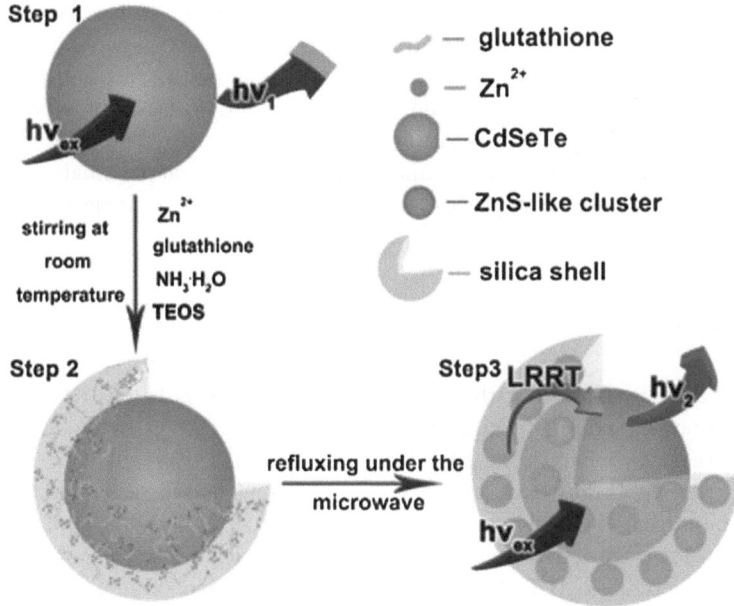

Fig. 2.1 Synthetic pathway for the preparation of CdSeTe@ZnS-SiO$_2$ QDs. Reproduced with permission from Ref. [12]. Copyright 2012, Royal Society of Chemistry

not new emerging member, such as silicon QDs, which appeared even as early as 1990. But due to their good biocompatibility, we summarized Si QDs in this section and metal nanoclusters are also included because of their excellent properties and wide applications as luminescent probes for biosensing and bioimaging.

2.2.1 Silicon Dots

Silicon has been known to be an indirect bandgap semiconductor with poor optical properties for a long time. It was until the 1990s when efficient light emission from silicon was reported by Canham [13] and quantum confinement to explain features of porous silicon absorption spectra was proposed by Lehman [14] that silicon nanocrystals attracted more and more interests of researchers. There are three distinct photoluminescence bands for Si QDs, one in the infrared, one in the red, and one in the blue light range. The strongest advantage of Si QDs as optical reporter lies in their good biocompatibility. Si QDs were claimed to be at least 10 times safer than Cd-based QDs under UV irradiation [15], and Canham even proposed nanoscale Si as a food additive [16]. Until now, numerous methods have been reported to produce colloidally and optically stable, water-dispersible Si QDs, incorporating a range of bottom–up and top–down approaches [17]. However, a key obstacle for their applications in bioimaging results from their oxidative degradation in the biological environment. For a solution, surface modification is necessary. Erogbogbo et al. prepared Si QDs through a nanoparticle synthesis, surface functionalization, PEGylated micelle encapsulation, and bioconjugation process. The obtained Si QDs could be used in multiple cancer-related in vivo applications, including tumor vasculature targeting, sentinel lymph node mapping, and multicolor NIR imaging in live mice, which showed great potentials of Si QDs as biocompatible fluorescent probes for both in vitro and in vivo imaging [18].

2.2.2 Carbon Dots

C-dots are a new class of carbon nanomaterials with sizes below 10 nm, which were first obtained during purification of single-walled carbon nanotubes through preparative electrophoresis in 2004 [19]. Since the discovery of their excellent optical property, C-dots have attracted wide attentions and displayed great potentials in biological applications. A special optical property of C-dots is that besides normal or down-converted photoluminescence, they were shown to possess excellent up-converted PL (UCPL), which enables the design of high-performance, complex catalyst systems based on C-dots for efficient utilization of the full spectrum of sunlight [20–23]. Additionally, C-dots can exhibit PL emission in the near-infrared (NIR) spectral region under NIR light excitation, which is particularly significant and useful for in vivo bionanotechnology because of the low autofluorescence and high tissue transparency in the NIR region [24, 25]. Except strong fluorescence, C-dots also own other properties such as electrochemical luminescence [26–28], photoinduced electron transfer property [29, 30], photocatalysis [22], optoelectronics [31, 32], which all extend their applications in various areas.

As a type of C-dots, the GQDs have also attracted a lot of interest from researchers over the past few decades because of their fascinating optical and electronic properties. As graphene is a zero-bandgap material, in principle, the bandgap of graphene can be tuned from 0 eV to that of benzene by varying their sizes [33, 34]. The 1D graphene sheets could be converted into 0D GQDs, which assume numerous novel chemical and physical properties due to the pronounced quantum confinement and edge effects [35, 36]. Although GQDs are considered as a member of C-dot family, there are still some differences between them [37]. The C-dots are either amorphous or crystalline, while GQDs possess graphene lattices inside the dots, regardless of the dot sizes [38]. Additionally, luminescent C-dots comprise discrete, quasi-spherical carbon nanoparticles with sizes below 10 nm, while GQDs are always defined as the graphene sheets with lateral dimensions than 100 nm in single, double, and few (3 to <10) layers [31, 34]. In general, the average sizes of GQDs are mostly below 10 nm, and up to now, the largest diameter of GQDs reported is 60 nm, which is dependent on the preparation methods [39]. Similar to C-dots, the properties of photoluminescence, good electron mobility and chemical stability, electrochemical luminescence, and photocatalyst of GQDs have been widely employed in the fabrication of numerous sensors and bioimaging

[37, 40]. Except Si QDs, C-dots, or GQDs, other kinds of cadmium-free QDs have also been well developed due to their good biocompatibility and excellent optical properties, such as InP [41], InP/ZnS [42], $CuInS_2/ZnS$ [43, 44], Ag_2Se [45], Ag_2S [46]. They all showed promising potentials in biological imaging applications.

2.2.3 Metal Nanoclusters

As a viable alternative to QDs, fluorescent metal nanoclusters, known as ultrasmall size, good biocompatibility, and excellent photostability, have become a new class of fluorescent labels for biological applications. Among them, Au and Ag nanoclusters attract much more attentions. Actually, at the beginning of the observation of photoluminescence from the noble metals, little attention was paid due to the extremely low quantum yield (QY) of 10^{-10} and much more interests have been attracted by researchers until the much enhanced QY reached to the range of 10^{-3} to 10^{-1} [47]. Until now, a lot of Au and Ag NCs stabilized with different scaffolds (protein, peptide, and oligonucleotide) have been developed and applied for the detection of thiol compounds [48], metal ions [49, 50], protein [51, 52], DNA [53], RNA [54] as well as intracellular and in vivo bioimaging. Dickson and coworkers successfully transferred poly (acrylic acid)-stabilized Ag NCs (PA-SCs) to anti-actin Ab/C12 and anti-α-tubulin/C12 conjugates to obtain fluorogenic silver cluster biolabels for cell surface labeling [55]. Wang et al. reported fluorescent Au NCs could be spontaneously biosynthesized by cancerous cell incubated with micromolar chloroauric acid solutions, a biocompatible molecular Au (III) species, which could not occur in noncancerous cells. They further realized in vivo self-bioimaging of tumors by subcutaneous injections of millimolar chloroauric acid solution near xenograft tumors of the nude mouse model of hepatocellular carcinoma or chronic myeloid leukemia. This opens up promising opportunities of fluorescent metal nanoclusters for in vivo bioimaging [56]. Specially, DNA-stabilized Ag NCs possess obvious advantage in DNA biosensing because of the easy assembly of DNA sequence. Werner and colleagues designed a nanocluster beacon to detect a DNA sequence related to the human *Braf* oncogene based on an interesting phenomenon that the red fluorescence of DNA-stabilized Ag NCs could be enhanced 500-fold when placed in proximity to guanine-rich DNA sequences [57].

2.3 Preparation and Functionalization

2.3.1 Cadmium-Based Quantum Dots

A lot of synthetic methods have been developed for QD preparation, which can be divided into two classifications: physical approach and chemical approach. Physical approach, which was mainly referred to epitaxial growth and/or nanoscale patterning, has been widely used to provide QDs predominantly by the combination of high-resolution electron beam lithography and subsequent etching. However, the production of defection formation, size nonuniformity, poor interface quality, and even damage to the bulk of the crystal itself became the main disadvantages of this method. In contrast, the size and composition of QDs prepared by chemical approach are easily controlled and the average size distribution varies within 5–10 %. The basic process of this method relies on the pyrolysis of organometallic and chalcogen precursors, where rapid nucleation followed by slower and steady growth is desired [58]. A typical protocol involved the following steps: Firstly, heat tri-*n*-octylphosphine oxide (TOPO) to a high temperature under argon or nitrogen atmosphere and then inject a hot solution containing the precursors to initiate rapid homogeneous nucleation. After removing the heat to lower the temperature of the reaction mixture quickly, the crystal growth continues for some time at a lower temperature [59]. Take the synthesis of cadmium selenide (CdSe) QDs, for example. Me_2Cd was chosen as the Cd source, and TOPSe was selected as chalcogen source due to their ease of preparation and good stability. The precursors were rapidly injected into the hot TOPO solution at 300 °C with vigorous stirring to produce a deep yellow/orange solution. Such rapid injection was accompanied by a sudden decrease in temperature to ~180 °C, and the nucleation was stopped. After restoring to heat the reaction flask to 230–260 °C, the growth of QDs continued and their sizes were further controlled by the reaction time [4]. However, Me_2Cd as the raw material was toxic, combustible, expensive, and unstable at room temperature. Meanwhile, the formed insoluble metallic precipitate after the injection of Me_2Cd into the hot TOPO also limited the wide use of this method. As a safer alternative, cadmium oxide (CdO) was chosen as an effective cadmium precursor first proposed by Peng et al. CdO, TOPO, and HPA/TDPA were loaded in a three-neck flask. At about 300 °C, reddish CdO powder was dissolved and generated a colorless homogenous solution. Then, the introduction of tellurium, selenium, and sulfur stock solutions yields high-quality quantum rods and dots of CdTe, CdSe, and CdS. This one-pot synthetic scheme brought a major step toward a green chemistry approach for synthesizing high-quality semiconductor nanocrystals [60, 61].

However, the applications of QDs in biological systems required they are water soluble. Therefore, numerous methods have been developed for creating hydrophilic QDs. One effective route is to exchange the hydrophobic layer of organic solvent with hydrophilic ligands such as thiol-containing molecules [62–66] and peptides [67], which was designated as "cap exchange." Another route is native surface modification, for example encapsulation by a layer of amphiphilic diblock [68, 69], triblock copolymers [70, 71], silica coating [72, 73], or phospholipid micelles [74, 75]. This two-step method for the hydrophilic QDs preparation broadened their biological applications greatly. However, direct synthesis of hydrophilic QDs in aqueous solution might be more favorable because of its simplicity, high reproducibility, and lower toxicity [76]. To achieve this aim, 3-mercaptopropionic acid (MPA) [77–79], 2-mercaptoethylamine acid (MA), thioglycolic acid (TGA) [80], and L-cysteine [81–83] are used as the stabilizers for one-step synthesis of hydrophilic QDs. Gao et al. prepared water-soluble CdTe QDs through the reaction between Cd^{2+} and NaHTe, with TGA as the stabilizer. They controlled the ratio of Cd^{2+}, NaHTe, and TGA and adjusted the pH value to 4.5–5.0. The combination between thiol group of TGA and Cd^{2+} promoted not only the hydrophilic stability but also the photoluminescent quantum yield [84, 85]. Besides, microwave-assisted green synthesis has been popular for the preparation of QDs. In comparison with conventional thermal techniques, microwave dielectric hearting has a few merits, such as fast heating and 1–2 orders of magnitude increase in the kinetics of the reaction rate. Specifically, microwave dielectric heating could realize the rapid and homogeneous growth of nanocrystals, which is extraordinarily beneficial for preparing high-quality NCs. With the help of microwave irradiation, CdTe, CdTe/ZnS, and CdTe/CdS/ZnS nanocrystals with high photoluminescent quantum yield and excellent photostability were synthesized successfully by He et al. [86–88].

2.3.2 Cadmium-Free Quantum Dots

Compared with cadmium-based QDs, available protocols for the synthesis of Si QDs are limited. As a whole, strategies for the preparation of Si QDs are generally composed of solution-phase-based methods [89–91], microemulsion synthesis [92], thermally induced disproportionation of solid hydrogen silsesquioxane in a reducing atmosphere, and so on [93]. Swihart's group successfully prepared water-dispersible Si QDs with blue, green, and yellow photoluminescence by the functionalization with acrylic acid in the presence of HF. However, as illustrated, these Si QDs still cannot satisfy the colloidal and spectral stability in biological environments [94]. Then, they further proposed a method for the preparation of water-dispersible and biocompatible Si QDs using phospholipid micelles. They were prepared by laser-driven pyrolysis of silane, followed by $HF-HNO_3$ etching, and the obtained Si QDs were dispersible in chloroform because of the surface functionalization of styrene, octadecene, or ethyl undecylenate. For water solubility, phospholipid micelles were then introduced and a hydrophilic shell with PEG groups was formed on the surface of Si QDs. Such micelle-encapsulated Si QDs displayed good application as biological luminescent probe in in vitro cell labeling [95]. Kauzlarich proposed a microwave-assisted reaction to produce hydrogen-terminated Si QDs. Two different methods were developed for the water-soluble Si QDs: hydrosilylation produced 3-aminopropenyl-terminated Si QDs, and a modified Stober process produced silica-encapsulated Si QDs. Both of them exhibited a maximum emission at 414 nm with intrinsic fluorescence quantum yield efficiencies of 15 and 23 %, respectively [96].

Synthetic methods for C-dots and GQDs are generally classified into two categories: top–down method and bottom–up method. Top–down methods commonly make use of laser ablation and electrochemical oxidation, where C-dots and GQDs are formed or "broken off" from a larger carbon structure and larger graphene sheets, respectively. Kang et al. developed an alkali-assisted electrochemical method to prepare C-dots with graphite rods as both anode and cathode and NaOH/EtOH as electrolyte. The judicious cutting of a graphite honeycomb layer into ultrasmall particles leads to tiny fragments of graphite, producing C-dots (Fig. 2.2a) [22]. Pan et al. prepared GQDs with bright blue photoluminescence by hydrothermal cutting of oxidized graphene sheets. The starting material, micrometer-sized ripped graphene sheets obtained by thermal reduction of graphene oxide sheets, was first oxidized in concentrated H_2SO_4 and HNO_3. After that, the hydrothermal treatment of the oxidized graphene sheets at 200 °C was performed, resulting in the dramatical decrease in the size of graphene sheets. Ultrafine GQDs were isolated by a dialysis process (Fig. 2.2b) [36]. Bottom–up approaches typically refer to solution chemistry methods during which C-dots and GQDs are formed from molecular precursors. For example, Li et al. reported a solution-chemistry-based approach to large, stable colloidal GQDs with uniform size and shape based on oxidative condensation reactions. The oxidation of polyphenylene dendritic precursors led to fused graphene moieties, which were then stabilized by multiple $2',4',6'$-trialkyl phenyl groups. The crowdedness on the edges of the graphene cores twists the substituted phenyl groups from the plane of the core, leading to alkyl chains closing the latter in all three dimensions and reducing face-to-face interaction between the graphenes to produce large, stable GQDs [97]. Puvvada et al. [98] prepared water-soluble C-dots through microwave-assisted pyrolysis of an aqueous solution of dextrin in the presence of sulfuric

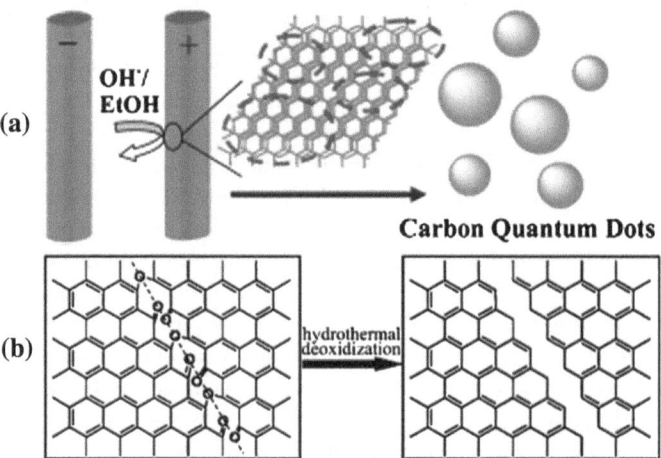

Fig. 2.2 a Schematic diagram of electrochemical fabrication of C-dots. Reproduced with permission from Ref. [22]. Copyright 2012, Royal Society of Chemistry. **b** Mechanism for the hydrothermal cutting of oxidized graphene sheets into GQDs. A mixed epoxy chain composed of epoxy and carbonyl pair groups (*left*) is converted into a complete cut (*right*) under the hydrothermal treatment. Reproduced with permission from Ref. [36]. Copyright 2010, Wiley

acid. It should be noted that the cage-opening of fullerene is another bottom–up approach for the synthesis of GQDs. Loh et al. successfully obtained GQDs by metal-catalyzed cage-opening of C_{60} [99].

2.3.3 Metal Nanoclusters

The metal nanoclusters are formed by the reduction of metal ions. However, such reduction of metal ions in aqueous solution always results in large nanoparticles rather than small NCs due to the tendency of NCs to aggregate [100]. Therefore, choosing suitable stabilizer to protect clusters from aggregating and enhancing their fluorescence is the key to obtain small and highly fluorescent metal NCs. Depending on the protection group, Au/Ag NCs are prepared mainly in six scaffolds: thiol-containing small molecules, dendrimers, polymers, DNA oligonucleotides, peptides, and proteins. Each kind of template has different roles in the synthesis of Au or Ag NCs. DNA oligonucleotides are widely employed in the preparation of fluorescent Ag NCs as good stabilizers because silver ions possess a high affinity to cytosine bases on single-stranded DNA. A serial of Ag NCs

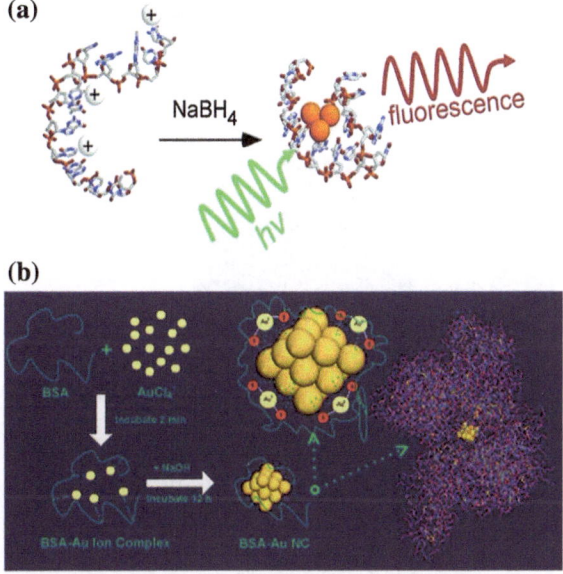

Fig. 2.3 a Schematic of the synthesis of DNA-stabilized Ag NCs. Reproduced with permission from Ref. [111]. Copyright 2004, American Chemical Society. **b** Schematic of the formation of Au NCs with BSA as the scaffold. Upon the addition of $HAuCl_4$ to the aqueous BSA solution under vigorous stirring, the protein molecules sequestered Au ions and entrapped them. After adjusting the pH to about 12 with NaOH, the Au ions were reduced by BSA to form Au NCs with red emissions in situ. Reproduced with permission from Ref. [103]. Copyright 2009, American Chemical Society

with different fluorescent emission wavelengths have been produced with various DNA sequences as the stabilizers [101]. Contrary to tremendous reports on DNA-templated Ag NCs, studies on the synthesis of luminescent Au NCs with DNA as the capping agents are scarce. Only Chen et al. reported that the atomically mono-disperse Au NCs could be obtained by etching gold particles with the help of amino acids, proteins, and DNA under sonication in water [102]. Unlike DNA oligonucleotides, proteins as the scaffolds offer more potentials in fluorescent Au NC formation. The first fluorescent protein-templated Au NCs were reported by Ying and coworkers in 2009. They developed a simple, green synthetic route for the preparation of Au NCs with red emissions based on the capability of bovine serum albumin (BSA) to sequester and reduce Au precursors [103]. Except BSA, other proteins such as lysozyme [104], transferrin [105], and HRP enzyme [106] have also been proved to act as efficient scaffolds for producing fluorescent Au NCs. In 2007, Dickson et al. reported a significant advance in producing fluorescent Ag NCs in vivo by ambient temperature photoactivation with nucleolin protein as the scaffold. Inspired by this, they further designed a short peptide incorporating the specific amino acids most prevalent in nucleolin and several cysteine groups to stabilize fluorescent Ag NCs directly in phosphate buffer [107]. Details about thiol-containing molecules, dendrimers, and polymers as the scaffold for the synthesis of Au and Ag NCs can be referred to other reviews [100, 108–110]. The main synthetic approaches for DNA-templated Ag NCs and BSA-stabilized Au NCs are illustrated in Fig. 2.3.

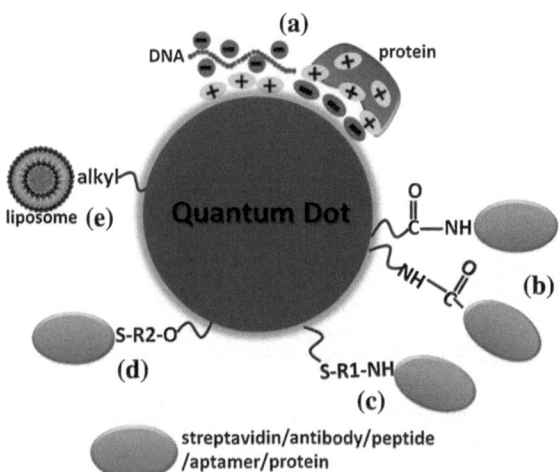

Fig. 2.4 Schematic presentation of various methods for the preparation of QD bioconjugates. **a** Electrostatic interaction between a positively charged protein and a negatively charged QD surface or between a negatively charged oligonucleotide and a positively charged QD surface. **b** Amide bond formation between carboxyl and amino groups by EDC/NHS chemistry. **c** Coupling between amine and thiol groups via the cross-linker SMCC. **d** Conjugation between hydroxyl and thiol groups. **e** Hydrophobic interactions between alkyl on QD surface and lipid or liposome

2.3.4 Quantum Dot Bioconjugation

Labeling of biomolecules, such as oligonucleotides, peptides, and proteins, to the QDs without disturbing the biological function of these molecules is usually required for their applications in biosensing. Commonly used approaches for QD bioconjugation with biological molecules include nonspecific adsorption, multivalent chelation, mercapto (-SH) exchange, and covalent linkage [112]. Several small molecules, such as oligonucleotides and serum albumins, are readily adsorbed to the surface of water-soluble QDs [113, 114]. However, such adsorption is influenced much by the ionic strength, pH, temperature, and the surface charge of the molecule. A more stable linkage is realized by covalently linking biomolecules to the functional groups of QDs using cross-linker molecules. Examples of covalent cross-linking methods include carbodiimide-mediated amide formation, active ester maleimide-mediated amine, and sulfhydryl coupling. Among them, carbodiimide-activated coupling between amine and carboxylic groups displays obvious advantage because most proteins contain primary amine and carboxylic acid groups and no further chemical functionalization is needed for QD conjugation [115, 116]. Figure 2.4 shown below presents a summary of various methods for the preparation of QD bioconjugates.

References

1. Ekimov AI, Onushchenko AA (1981) Quantum size effect in three-dimensional microscopic semiconductor crystals. JETP Lett 34(6):345–349
2. Brus L (1984) Electron-electron and electron-hole interactions in small semiconductor crystallites: the size dependence of the lowest excited electronic state. J Chem Phys 80:4403–4409
3. Brus L (1986) Electronic wave functions in semiconductor clusters: experiment and theory. J Phys Chem 90(12):2555–2560
4. Murray CB, Norris DJ, Bawendi MG (1993) Synthesis and characterization of nearly monodisperse CdE (E = sulfur, selenium, tellurium) semiconductor nanocrystallites. J Am Chem Soc 115(19):8706–8715
5. Hines MA, Guyot-Sionnest P (1996) Synthesis and characterization of strongly luminescing ZnS-capped CdSe nanocrystals. J Phys Chem 100(2):468–471
6. Dabbousi BO, Rodriguez-Viejo J, Mikulec FV, Heine JR, Mattoussi H, Ober R, Jensen KF, Bawendi MG (1997) (CdSe)ZnS core-shell quantum dots: synthesis and characterization of a size series of highly luminescent nanocrystallites. J Phys Chem B 101(46):9463–9475
7. Peng X, Schlamp MC, Kadavanich AV, Alivisatos AP (1997) Epitaxial growth of highly luminescent CdSe/CdS core/shell nanocrystals with photostability and electronic accessibility. J Am Chem Soc 119(30):7019–7029
8. Reiss P, Bleuse J, Pron A (2002) Highly luminescent CdSe/ZnSe core/shell nanocrystals of low size dispersion. Nano Lett 2(7):781–784
9. He Y, Lu HT, Sai LM (2006) Microwave-assisted growth and characterization of water-dispersed CdTe/CdS core-shell nanocrystals with high photoluminescence. J Phys Chem B 110(27):13370–13374
10. Zhao D, He Z, Chan WH, Choi MMF (2009) Synthesis and characterization of high-quality water-soluble near-infrared-emitting CdTe/CdS quantum dots capped by N-Acetyl-L-cysteine via hydrothermal method. J Phys Chem C 113(4):1293–1300

11. He Y, Lu HT, Sai LM, Su YY, Hu M, Fan CH, Huang W, Wang LH (2008) Microwave synthesis of water-dispersed CdTe/CdS/ZnS core-shell-shell quantum dots with excellent photostability and biocompatibility. Adv Mater 20(18):3416–3421

12. Shen YY, Li LL, Lu Q, Ji J, Fei R, Zhang JR, Abdel-Halim ES, Zhu J-J (2012) Microwave-assisted synthesis of highly luminescent CdSeTe@ZnS-SiO$_2$ quantum dots and their application in the detection of Cu(II). Chem Commun 48:2222–2224

13. Canham LT (1990) Silicon quantum wire array fabrication by electrochemical and chemical dissolution of wafers. Appl Phys Lett 57:1046–1048

14. Lehmann V, Gosele U (1991) Porous silicon formation: a quantum wire effect. Appl Phys Lett 58:656–658

15. Fujioka K, Hlruoka M, Sato K, Manabe N, Mlyasaka R, Hanada S, Hoshino A, Tilley RD, Manome Y, Hirakuri K, Yamamoto K (2008) Luminescent passive-oxidized silicon quantum dots as biological staining labels and their cytotoxicity effects as high concentration. Nanotechnology 19(41):415102

16. Canham LT (2007) Nanoscale semiconducting silicon as a nutritional food additive. Nanotechnology 18(18):185704–185709

17. Bruhn B (2012) Fabrication and characterization of single luminescing quantum dots from 1D silicon nanostructures. Doctoral Thesis

18. Erogbogbo F, Yong K-T, Roy I, Hu R, Law W-C, Zhao WW, Ding H, Wu F, Kumar R, Swihart MT, Prasad PN (2011) In vivo targeted cancer imaging, sentinel lymph node mapping and multi-channel imaging with biocompatible silicon nanocrystals. ACS Nano 5(1):413–423

19. Xu XY, Ray R, Gu YL, Ploehn HJ, Gearheart L, Raker K, Scrivens WA (2004) Electrophoretic analysis and purification of fluorescent single-walled carbon nanotube fragments. J Am Chem Soc 126(40):12736–12737

20. Shen JH, Zhu YH, Yang XL, Li CZ (2012) Graphene quantum dots: emergent nanolights for bioimaging, sensors, catalysis and photovoltaic devices. Chem Commun 48:3686–3699

21. Cao L, Wang X, Meziani MJ, Lu F, Wang H, Luo PG, Lin Y, Harruff BA, Veca LM, Murray D, Xie SY, Sun YP (2007) Carbon dots for multiphoton bioimaging. J Am Chem Soc 129(37):11318–11319

22. Li HT, Kang ZH, Liu Y, Lee S-T (2012) Carbon nanodots: synthesis, properties and applications. J Mater Chem 22:24230–24253

23. Ming H, Ma Z, Liu Y, Pan KM, Yu H, Wang F, Kang ZH (2012) Large scale electrochemical synthesis of high quality carbon nanodots and their photocatalytic property. Dalton Trans 41:9526–9531

24. Lim SF, Riehn R, Ryu WS, Khanarian N, Tung CK, Tank D, Austin RH (2006) In vivo and scanning electron microscopy imaging of upconverting nanophosphors in *caenorhabditis elegans*. Nano Lett 6(2):169–174

25. Tang LB, Ji RB, Cao XK, Lin JY, Jiang HX, Li XM, Teng KS, Luk CM, Zeng SJ, Hao JH, Lau SP (2012) Deep ultraviolet photoluminescence of water-soluble self-passivated graphene quantum dots. ACS Nano 6(6):5102–5110

26. Ding ZF, Quinn BM, Haram SK, Pell LE, Korgel BA, Bard AJ (2002) Electrochemistry and electrogenerated chemiluminescence from silicon nanocrystal quantum dots. Science 296:1293–1297

27. Zhu H, Wang XL, Li YL, Wang ZJ, Yang F, Yang XR (2009) Microwave synthesis of fluorescent carbon nanoparticles with electrochemiluminescence properties. Chem Commun 34:5118–5120

28. Zhou JG, Booker C, Li RY, Sun XL, Sham TK, Ding ZF (2010) Electrochemistry and electrochemiluminescence study of blue luminescent carbon nanocrystals. Chem Phys Lett 493:296–298

29. Zhang HC, Huang H, Ming H, Li HT, Zhang LL, Liu Y, Kang ZH (2012) Carbon quantum dots/Ag$_3$PO$_4$ complex photocatalysts with enhanced photocatalytic activity and stability under visible light. J Mater Chem 22:10501–10506

30. Wang X, Cao L, Lu FS, Meziani MJ, Li H, Qi G, Zhou B, Harruff BA, Kermarrec F, Sun YP (2009) Photoinduced electron transfers with carbon dots. Chem Commun 25:3774–3776

31. Ponomarenko LA, Schedin F, Katsnelson MI, Yang R, Hill EW, Novoselov KS, Geim AK (2008) Chaotic dirac billiard in graphene quantum dots. Science 320:356–358

32. Girit CO, Meyer JC, Erni R, Rossell MD, Kisielowski C, Yang L, Park CH, Crommie MF, Cohen ML, Louie SG, Zettl A (2009) Graphene at the edge: stability and dynamics. Science 323:1705–1708

33. Yan X, Cui X, Li B, Li LS (2010) Large, solution-processable graphene quantum dots as light absorbers for photovoltaics. Nano Lett 10:1869–1873

34. Shen J, Zhu Y, Yang X, Zong J, Zhang J, Li C (2012) One-pot hydrothermal synthesis of graphene quantum dots surface-passivated by polyethylene glycol and their photoelectric conversion under near-infrared light. New J Chem 36:97–101

35. Zhou X, Zhang Y, Wang C, Wu X, Yang Y, Zheng B, Wu H, Guo S, Zhang J (2012) Photo-fenton reaction of graphene oxide: a new strategy to prepare graphene quantum dots for DNA cleavage. ACS Nano 6:6592–6599

36. Pan DY, Zhang JC, Li Z, Wu MH (2010) Hydrothermal route for cutting graphene sheets into blue-luminescent graphene quantum dots. Adv Mater 22(6):734–738

37. Li LL, Wu GH, Yang GH, Peng J, Zhao JW, Zhu J-J (2013) Focusing on luminescent graphene quantum dots: current status and future perspectives. Nanoscale 5:4015–4039

38. Baker SN, Baker GA (2010) Luminescent carbon nanodots: emergent nanolights. Angew Chem Int Ed 49(38):6726–6744

39. Liu R, Wu D, Feng X, Muellen K (2011) Bottom–up fabrication of photoluminescent graphene quantum dots with uniform morphology. J Am Chem Soc 133:15221–15223

40. Shen JH, Zhu YH, Yang XL, Li CZ (2012) Graphene quantum dots: emergent nanolights for bioimaging, sensors, catalysis and photovoltaic devices. Chem Commun 48:3686–3699

41. Yong KT, Ding H, Roy I, Law W-C, Bergey EJ, Maitra A, Prasad PN (2009) Imaging pancreatic cancer using bioconjugated InP quantum dots. ACS Nano 3(3):502–510

42. Tamang S, Beaune G, Texier I, Reiss P (2011) Aqueous phase transfer of InP/ZnS nanocrystals conserving fluorescence and high colloidal stability. ACS Nano 5(12):9392–9402

43. Chen YY, Li SJ, Huang LJ, Pan DC (2013) Green and facile synthesis of water-soluble Cu-In-S/ZnS core/shell quantum dots. Inorg Chem 52:7819–7821

44. Li L, Daou TJ, Texier I, Chi TTK, Liem NQ, Reiss P (2009) Highly luminescent CuInS$_2$/ZnS core/shell nanocrystals: cadmium-free quantum dots for in vivo imaging. Chem Mater 21:2422–2429

45. Gu Y-P, Cui R, Zhang Z-L, Xie Z-X, Pang D-W (2012) Ultrasmall near-infrared Ag$_2$Se quantum dots with tunable fluorescence for in vivo imaging. J Am Chem Soc 134(1):79–82

46. Hong GS, Robinson JT, Zhang YJ, Diao S, Antaris AL, Wang QB, Dai HJ (2012) In vivo fluorescence imaging with Ag$_2$S quantum dots in the second near-infrared region. Angew Chem Int Ed 51(39):9818–9821

47. Mooradian A (1969) Photoluminescence of metals. Phys Rev Lett 22(5):185–187

48. Huang ZZ, Pu F, Lin YH, Ren JS, Qu XG (2011) Modulating DNA-templated silver nanoclusters for fluorescence turn-on detection of thiol compounds. Chem Commun 47:3487–3489

49. Su YT, Lan GY, Chen WY, Chang HT (2010) Detection of copper ions through recovery of the fluorescence of DNA templated copper/silver nanoclusters in the presence of mercaptopropionic acid. Anal Chem 82:8566–8572

50. Lan GY, Huang CC, Chang HT (2010) Silver nanoclusters as fluorescent probes for selective and sensitive detection of copper ions. Chem Commun 46:1257–1259

51. Sharma J, Yeh HC, Yoo H, Werner JH, Martinez JS (2011) Silver nanocluster aptamers: in situ generation of intrinsically fluorescent recognition ligands for protein detection. Chem Commun 47:2294–2296

52. Li JJ, Zhong XQ, Zhang HQ, Le XC, Zhu J-J (2012) Binding-induced fluorescence turn-on assay using aptamer-functionalized silver nanocluster DNA probes. Anal Chem 84:5170–5174

53. Yeh H-C, Sharma J, Shih I-M, Vu DM, Martinez JS, Werner JH (2012) A fluorescence light-up Ag nanocluster probe that discriminates single-nucleotide variants by emission color. J Am Chem Soc 134:11550–11558

54. Dong HF, Jin S, Ju HX, Hao KH, Xu L-P, Lu HT, Zhang XJ (2012) Trace and label-free microRNA detection using oligonucleotide encapsulated silver nanoclusters as probes. Anal Chem 84:8670–8674

55. Yu JH, Choi S, Dickson RM (2009) Shuttle-based fluorogenic silver-cluster biolabels. Angew Chem Int Ed 48(2):318–320

56. Wang JL, Zhang G, Li QW, Jiang H, Liu CY, Amatore C, Wang XM (2013) In vivo self-bio-imaging of tumors through in situ biosynthesized fluorescent gold nanoclusters. Sci Rep 3:1157–1162

57. Yeh H-C, Sharma J, Han JJ, Martinez JS, Werner JH (2010) A DNA-silver nanocluster probe that fluoresces upon hybridization. Nano Lett 10:3106–3110

58. Drbohlavova J, Adam V, Kizek R, Hubalek J (2009) Quantum dots-characterization, preparation and usage in biological systems. Int J Mol Sci 10(2):656–673

59. Algar WR, Krull UJ (2009) Biosensing using nanomaterials. In: Merkoci A (ed) Quantum dots for the development of optical biosensors based on fluorescence, 7th edn. Wiley, New York

60. Peng ZA, Peng XG (2001) Formation of high-quality CdTe, CdSe, and CdS nanocrystals using CdO as precursor. J Am Chem Soc 123(1):183–184

61. Qu LH, Peng XG (2002) Control of photoluminescence properties of CdSe nanocrystals in growth. J Am Chem Soc 124(9):2049–2055

62. Chan WC, Nie SM (1998) Quantum dot bioconjugates for ultrasensitive nonisotopic detection. Science 281:2016–2018

63. Guo WZ, Li JJ, Wang YA, Peng XG (2003) Conjugation chemistry and bioapplications of semiconductor box nanocrystals prepared via dendrimer bridging. Chem Mater 15:3125–3133

64. Jaiswal JK, Mattoussi H, Mauro JM, Simon SM (2003) Long-term multiple color imaging of live cells using quantum dot bioconjugates. Nat Biotechnol 21(1):47–51

65. Zeng Q, Kong X, Sun Y, Zhang Y, Tu L, Zhao J, Zhang H (2008) Synthesis and optical properties of type II CdTe/CdS core/shell quantum dots in aqueous solution via successive ion layer adsorption and reaction. J Phys Chem C 112:8587–8593

66. Zeng R, Zhang T, Liu J, Hu S, Wan Q, Liu X, Peng Z, Zou B (2009) Aqueous synthesis of type-II CdTe/CdSe core-shell quantum dots for fluorescent probe labeling tumor cells. Nanotechnology 20:095102

67. Pinaud F, King D, Moore HP, Weiss S (2004) Bioactivation and cell targeting of semiconductor CdSe/ZnS nanocrystals with phytochelatin-related peptides. J Am Chem Soc 126:6115–6123

68. Liu W, He ZK, Liang JG, Zhu YL, Xu HB, Yang XL (2008) Preparation and characterization of novel fluorescent nanocomposite particles: CdSe/ZnS core-shell quantum dots loaded solid lipid nanoparticles. J Biomed Mater Res Part A 84:A1018–A1025

69. Wu XY, Liu HJ, Liu JQ, Haley KN, Treadway JA, Larson JP, Ge NF, Peale F, Bruchez MP (2003) Immunofluorescent labeling of cancer marker Her2 and other cellular targets with semiconductor quantum dots. Nat Biotechnol 21(1):41–46

70. Zhang BB, Cheng J, Li DN, Liu XH, Ma GP, Chang J (2008) A novel method to make hydrophilic quantum dots and its application on biodetection. Mater Sci Eng B-Adv Funct Solid-State Mater 149:87–92

71. Gao XH, Cui YY, Levenson RM, Chung LWK, Nie SM (2004) In vivo caner targeting and imaging with semiconductor quantum dots. Nat Biotechnol 22(8):969–976

72. Koole R, van Schooneveld MM, Hilhorst J, Donega CD, Hart DC, van Blaaderen A, Vanmaekelbergh D, Meijerink A (2008) On the incorporation mechanism of hydrophobic quantum dots in silica spheres by a reverse microemulsion method. Chem Mat 20:2503–2512

73. Bruchez MJ, Moronne M, Gin P, Weiss S, Alivisatos AP (1998) Semiconductor nanocrystals as fluorescent biological labels. Science 281:2013–2016

74. Dubertret B, Skourides P, Norris DJ, Noireaux V, Brivanlou AH, Libchaber A (2002) In vivo imaging of quantum dots encapsulated in phospholipid micelles. Science 298:1759–1762

75. Michalet X, Pinaud FF, Bentolila LA, Tsay JM, Doose S, Li JJ, Sundaresan G, Wu AM, Gambhir SS, Weiss S (2005) Quantum dots for live cells, in vivo imaging, and diagnostics. Science 307:538–544

76. Gaponik N, Talapin DV, Rogach AL, Hoppe K, Shevchenko EV, Kornowski A, Eychmuller A, Weller H (2002) Thiol-capping of CdTe nanocrystals: an alternative to organometallic synthetic routes. J Phys Chem B 106:7177–7185

77. Asha JM, Arunkumar K, Rajalingam R, Mahmmoud Sayed Abd E-S (2011) Photoinduced interaction of MPA-capped CdTe quantum dots with denatured bovine serum albumin. Nanosci Nanotechnol Lett 3(2):125–130

78. Haque MM, Im HY, Seo JE, Hasan M, Woo K, Kwon OS (2012) Acute toxicity and tissue distribution of CdSe/CdS-MPA quantum dots after repeated intraperitoneal injection to mice. J Appl Toxicol 33(9):940–950

79. Emamdoust A, Shayesteh SF, Marandi M (2013) Synthesis and characterization of aqueous MPA-capped CdS–ZnS core-shell quantum dots. Pramana J Phys 80(4):713–721

80. Gallaqher S, Comby S, Wojdyla M, Gunnlauqsson T, Kelly JM, Gun'ko YK, Clark IP, Greetham GM, Towrie M, Quinn SJ (2013) Efficient quenching of TGA-capped CdTe quantum dot emission by a surface-coordinated europium(III) cyclen complex. Inorg Chem 52(8):4133–4135

81. Kumar P, Kumar P, Bharadwaj LM, Paul AK, Sharma SC, Kush P, Deep A (2013) Aqueous synthesis of L-cysteine stabilized water-dispersible CdS: Mn quantum dots for biosensing applications. BioNanoSci 3(2):95–101

82. Zhang YH, Zhang HS, Guo XF, Wang H (2008) L-Cysteine-coated CdSe/CdS core-shell quantum dots as selective fluorescence probe for copper(II) determination. Microchem J 89(2):142–147

83. Zhang H, Sun P, Liu C, Gao H, Xu L, Fang J, Wang M, Liu J, Xu S (2011) L-Cysteine capped CdTe-CdS core-shell quantum dots: preparation, characterization and immuno-labeling of HeLa cells. Luminescence 26(2):86–92

84. Gao M, Rogach AL, Kornowski A (1998) Strongly photoluminescent CdTe nanocrystals by proper surface modification. J Phys Chem B 102:8360–8363

85. Zhang H, Zhou Z, Yang B (2003) The influence of carboxyl groups on the photoluminescence of mercaptocarboxylic acid-stabilized CdTe nanoparticles. J Phys Chem B 107:8–13

86. He Y, Sai L-M, Lu H-T, Hu M, Lai W-Y, Fan Q-L, Wang L-H, Huang W (2007) Microwave-assisted synthesis of water-dispersed CdTe nanocrystals with high luminescent efficiency and narrow size distribution. Chem Mater 19:359–365

87. He Y, Lu H-T, Sai L-M, Lai W-Y, Fan Q-L, Wang L-H, Huang W (2006) Microwave-assisted growth and characterization of water-dispersed CdTe/CdS core-shell nanocrystals with high photoluminescence. J Phys Chem B 110:13370–13374

88. He Y, Lu H-T, Sai L-M, Su Y-Y, Hu M, Fan C-H, Huang W, Wang L-H (2008) Microwave synthesis of water-dispersed CdTe/CdS/ZnS core-shell-shell quantum dots with excellent photostability and biocompatibility. Adv Mater 20(18):3416–3421

89. Warner JH, Hoshino A, Yamamoto K, Tilley RD (2005) Water-soluble photoluminescent silicon quantum dots. Angew Chem Int Ed 44(29):4550–4554

90. Neiner D, Chiu HW, Kauzlarich SM (2006) Low-temperature solution route to macroscopic amounts of hydrogen terminated silicon nanoparticles. J Am Chem Soc 128:11016–11017

91. Neiner D, Kauzlarich SM (2010) Hydrogen-capped silicon nanoparticles as a potential hydrogen storage material: synthesis, characterization, and hydrogen release. Chem Mater 22:487–493

92. Wilcoxon JP, Samara GA (1999) Tailorable, visible light emission from silicon nanocrystals. Appl Phys Lett 74:3164–3166

93. Hessel CM, Henderson EJ, Veinot JGG (2007) An investigation of the formation and growth of oxide-embedded silicon nanocrystals in hydrogen silsesquioxane-derived nanocomposites. J Phys Chem C 111:6956–6961

94. Sato S, Swihart MT (2006) Propionic-acid terminated silicon nanoparticles: synthesis and optical characterization. Chem Mater 18:4083–4088

95. Erogbogbo F, Yong K-T, Roy I, Xu G, Prasad PN, Swihart MT (2008) Biocompatible luminescent silicon quantum dots for imaging of cancer cells. ACS Nano 2(5):873–878

96. Atkins TM, Thibert A, Larsen DS, Dey S, Browning ND, Kauzlarich SM (2011) Femtosecond ligand/core dynamics of microwave-assisted synthesized silicon quantum dots in aqueous solution. J Am Chem Soc 133(51):20664–20667

97. Yan X, Cui X, Li L-S (2010) Synthesis of large, stable colloidal graphene quantum dots with tunable size. J Am Chem Soc 132(17):5944–5945

98. Puvvada N, Kumar BNP, Konar S, Kalita H, Mandal M, Pathak A (2012) Synthesis of biocompatible multicolor luminescent carbon dots for bioimaging applications. Sci Technol Adv Mater 12:045008

99. Lu J, Yeo PSE, Gan CK, Wu P, Loh KP (2011) Transforming C_{60} molecules into graphene quantum dots. Nat Nanotechnol 6(4):247–252

100. Diez I, Ras RHA (2011) Fluorescent silver nanoclusters. Nanoscale 3:1963–1970

101. Richards CI, Choi S, Hsiang J-C, Vosch YAT, Bongiorno A, Tzeng Y-L, Dickson RM (2008) Oligonucleotide-stabilized Ag nanocluster fluorophores. J Am Chem Soc 130:5038–5039

102. Zhou R, Shi M, Chen X, Wang M, Chen H (2009) Atomically monodispersed and fluorescent sub-nanometer gold clusters created by biomolecule-assisted etching of nanometer-sized gold particles and rods. Chem Eur J 15(19):4944–4951

103. Xie J, Zheng Y, Ying JY (2009) Protein-directed synthesis of highly fluorescent gold nanoclusters. J Am Chem Soc 131:888–889

104. Wei H, Wang Z, Yang L, Tian S, Hou C, Lu Y (2010) Lysozyme-stabilized gold fluorescent cluster: synthesis and application as Hg^{2+} sensor. Analyst 135:1406–1410

105. Guevel XL, Daum N, Schneider M (2011) Synthesis and characterization of human transferrin-stabilized gold nanoclusters. Nanotechnology 22:275103

106. Wen F, Dong Y, Feng L, Wang S, Zhang S, Zhang X (2011) Horseradish peroxidase functionalized fluorescent gold nanoclusters for hydrogen peroxide sensing. Anal Chem 83:1193–1196

107. Yu J, Patel SA, Dickson RM (2007) In vitro and intracellular production of peptide-encapsulated fluorescent silver nanoclusters. Angew Chem Int Ed 46(12):2028–2030

108. Shang L, Dong SJ, Nienhaus GU (2011) Ultra-small fluorescent metal nanoclusters: synthesis and biological applications. Nano Today 6:401–418

109. Choi S, Dickson RM, Yu JH (2012) Developing luminescent silver nanodots for biological applications. Chem Soc Rev 41:1867–1891

110. Obliosca JM, Liu C, Yeh H-C (2013) Fluorescent silver nanoclusters as DNA probes. Nanoscale 5(18):8443–8461

111. Petty JT, Zheng J, Hud NV, Dickson RM (2004) DNA-templated Ag nanocluster formation. J Am Chem Soc 126:5207–5212

112. Alivisatos P, Gu W, Larabell C (2005) Quantum dots as fluorescent probes. Ann Rev Biomed Eng 7:55–76

113. Xing Y, Rao JH (2008) Quantum dot bioconjugates for in vitro diagnostics & in vivo imaging. Cancer Biomarkers 4:307–319

114. Lakowicz JR, Gryczynski I, Gryczynski Z, Nowaczk K, Murphy CJ (2000) Time-resolved spectral observations of cadmium-enriched cadmium sulfide nanoparticles and the effects of DNA oligomer binding. Anal Biochem 280:128–136

115. Mahtab R, Harden HH, Murphy CJ (2000) Temperature- and salt- dependent binding of long DNA to protein-sized quantum dots: thermodynamics of "inorganic protein"-DNA interactions. J Am Chem Soc 122:14–17

116. Gao XH, Yang L, Petros JA, Marshall FF, Simons JW, Nie SM (2005) In vivo molecular and cellular imaging with quantum dots. Curr Opin Biotechnol 16:63–72

Chapter 3
Quantum Dot-Fluorescence-Based Biosensing

Abstract Since the emergence of quantum dots (QDs), their excellent fluorescent properties have been widely used in the fabrication of biological sensors for various analytes, such as metal ions, environmental samples, protein, deoxyribonucleic acid (DNA), and ribonucleic acid (RNA). More importantly, some of these sensors can realize not only fixed cell labeling, imaging of live cell dynamics, in situ tissue profiling, but also in vivo animal imaging. A lot of reviews have well summarized these in vitro diagnostic applications and in vivo imaging and sensing applications of QDs. In this chapter, we mainly focus on QDs-fluorescence-based biosensing in DNA, RNA, and DNA microarrays. The introduction of QDs in this aspect promoted the sensitivity, stability, and diversity of DNA and RNA detection obviously.

Keywords Quantum dots • DNA biosensor • Fluorescence • RNA analysis • DNA microarray

3.1 QDs for DNA Analysis

3.1.1 Main Types for DNA Detection

As illustrated in the introduction part, deoxyribonucleic acid (DNA) biosensing has important significance because of the useful information they possess for disease diagnosis. One simple type of DNA detection systems is based on the hybridization between a DNA target and its complementary probe, where DNA target is commonly directly labeled with QDs. However, such modification of DNA target has less feasibility in practice. For better applications, other types of DNA biosensors such as sandwich structure, competitive system, and molecular beacon (MB) have been studied a lot during the past decades (Fig. 3.1). Fluorescence resonance energy transfer (FRET) has been widely employed in these kinds of DNA biosensors. FRET is a nonradiative process whereby an excited-state donor D (usually a fluorophore) transfers energy to a

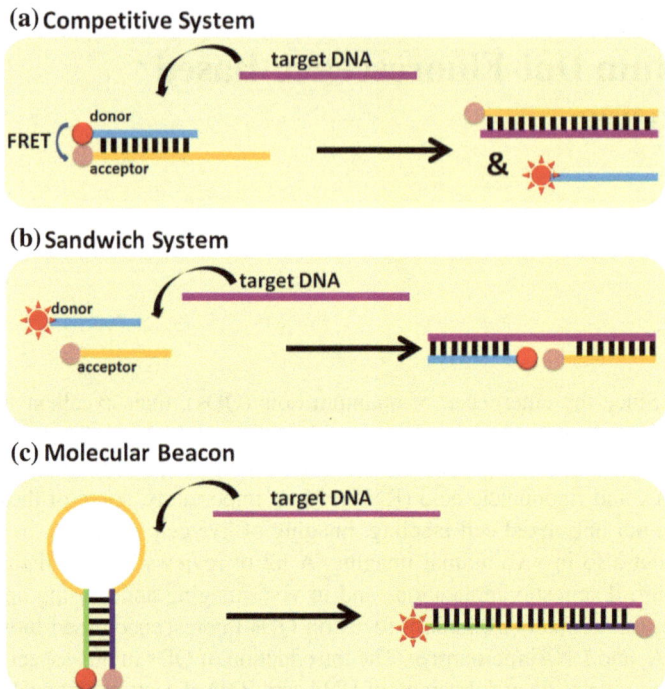

Fig. 3.1 Main types for DNA detection. **a** competitive system, **b** sandwich system, and **c** molecular beacon structure

proximal ground-state acceptor A through long-range dipole–dipole interactions [1–3]. The acceptor must absorb energy at the emission wavelength(s) of the donor, but does not necessarily have to remit the energy fluorescently itself. The rate of energy transfer is highly dependent on many factors, such as the extent of spectral overlap, the relative orientation of the transition dipoles, and, most importantly, the distance between the donor and acceptor molecules [4, 5]. FRET is very appealing for bioanalysis because of its simpleness of building ratiometric fluorescent systems. At the beginning, the FRET-based sensing systems were designed with organic dye as donors. With the appearance and development of QDs, more and more QD-based FRET DNA biosensors have emerged. The wide use of QDs as the donors in FRET can be ascribed not only to their high fluorescence quantum yield, strong resistance to photobleaching, but also to their broad excitation wavelengths and narrow and symmetric size-tunable emission spectra. These characteristic properties make their great promise in FRET assays [6]. This method avoided the numerous problems linked to DNA intercalating dyes commonly used for DNA imaging (photobleaching, photoinduced cleavage, and modification of the DNA properties). Krull et al. developed a multiplexed solid-phase nucleic acid hybridization assay on a paper-based platform using multicolor immobilized QDs as donors in FRET, which could reach a detection limit of 90 fmol and an upper limit of dynamic range of 3.5 pmol. As shown in Fig. 3.2, the surface of paper

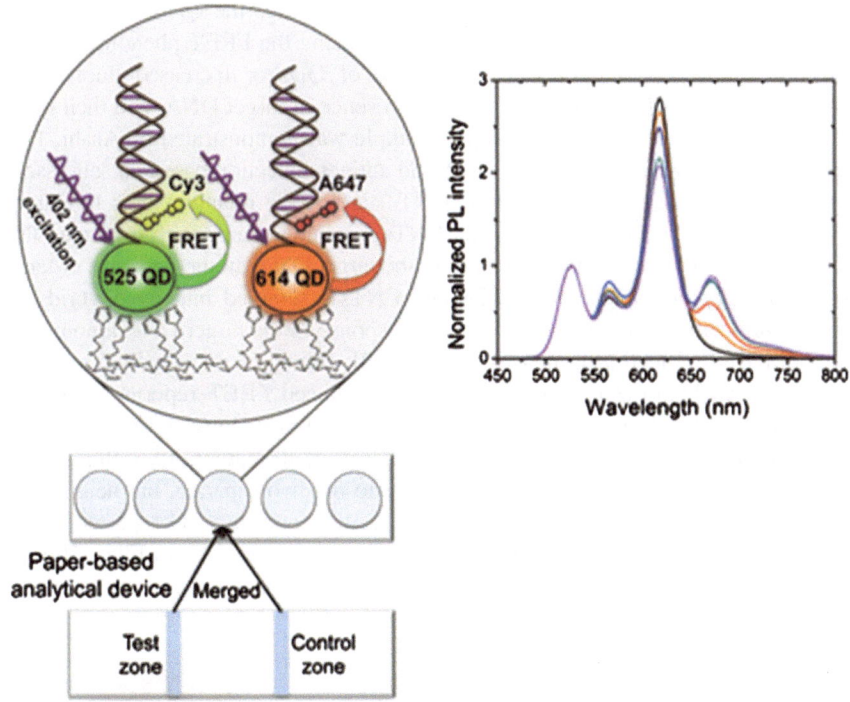

Fig. 3.2 Design of the paper-based solid-phase multiplexed nucleic acid hybridization assay using multicolor immobilized QDs as donors in FRET. Hybridization with Cy3-labeled SMN1 and A647-labeled target oligonucleotides provided the proximity for FRET-sensitized emission from Cy3 and A647 dyes. Reproduced with permission from Ref. [7]. Copyright 2013, American Chemical Society

was modified with imidazole groups to immobilize two types of QDs–DNA probes. Green-emitting QDs (gQDs) and red-emitting QDs (rQDs) served as donors with Cy3 and Alexa Fluor 647 (A647) as acceptors. The gQD/Cy3 FRET pair served as an internal standard and the rQD/A647 FRET pair served as a detection channel, combining the control and analytical test zones in one physical location. Hybridization of dye-labeled oligonucleotide targets provided the proximity for FRET-sensitized emission from the acceptor dyes, which serves as an analytical signal [7]. Recently, new emerging zero-bandgap carbon nanomaterials, graphene, and graphene oxide (GO) have been widely used as energy acceptors combining with QDs as donors in DNA sensing. He et al. prepared DNA–CdTe by a one-pot method and developed a biosensor based on FRET from the DNA-functionalized CdTe nanocrystals to graphene for the detection of the hepatitis B virus (HBV) surface antigen gene [8].

In a typical sandwich structure, two specific DNA probes are required, i.e., reporter and capture probe. These two probes are labeled with QDs as donor and fluorophore or quencher as acceptor, respectively. They are partly complementary to the target DNA sequence. Thus, in the presence of target DNA, it becomes sandwiched by reporter and capture probes, which brings the QD donor and fluorophore or quencher acceptor close proximity causing the FRET phenomenon. The detection of decreased fluorescence emission of QDs or increased fluorescence emission intensity of acceptor indicates the presence of target DNA, and their quantification can also be realized. A simple example was demonstrated by Asahi. They developed a sandwich structure for DNA and antigen detection based on self-assembly of multiwalled carbon nanotubes (CNTs) and CdSe quantum dots (QDs) via oligonucleotide hybridization. QDs and CNTs were conjugated with different DNA oligos, which were both complementary to the target DNA. In the presence of target complementary oligonucleotides, QDs and CNTs assembled into nanohybrids via DNA hybridization bringing a fluorescence response to the target DNA amount and a 0.2-pM DNA detection limit was achieved [9]. Huang et al. described a nucleic acid sandwich hybridization assay with a QD-induced FRET reporter system, as shown in Fig. 3.3. Hemagglutinin H5 sequences (60-mer DNA and 630-nt cDNA fragment) of avian influenza viruses were chosen as the target. Two oligonucleotides (16 mers and 18 mers) that complement to the two separate, but neighboring regions of the target sequence were designed as the capturing and reporter probes, respectively. They were conjugated with QD655 (donor) and Alexa Fluor 660 dye (acceptor) at first. The sandwich hybridization occurred once the existence of target

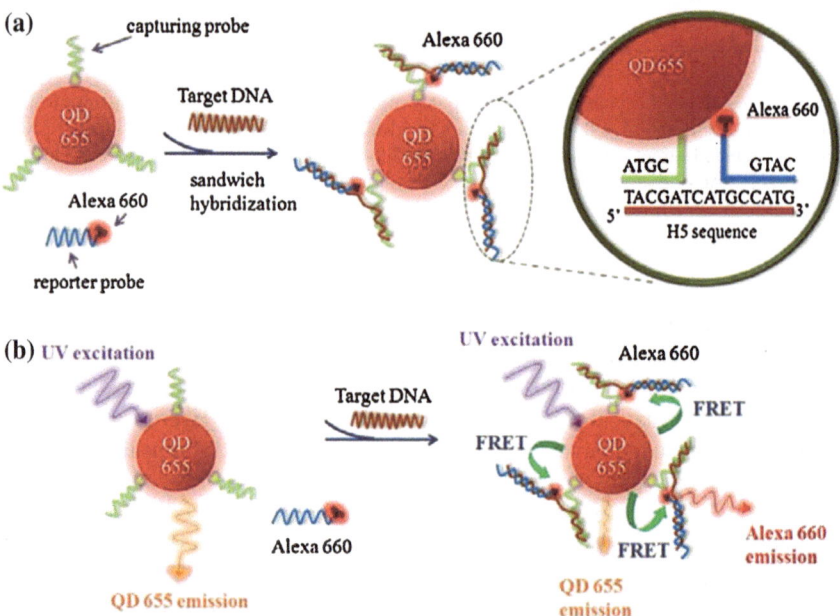

Fig. 3.3 Schematic illustration of the sandwich hybridization assay with a QD-induced FRET reporter system for H5 target DNA detection. **a** Sandwich hybridization with H5 sequence (target) by the capturing probes conjugated on QD655 (FRET donor) and the reporter probes labeled with Alexa Fluor 660 (FRET acceptor); **b** FRET emission shift before and after the sandwich hybridization. Reproduced with permission from Ref. [10]. Copyright 2012, MDPI AG

sequence causing a FRET signal response, which was monitored by a homemade optical sensor comprising a single 400-nm UV light-emitting diode (LED), optical fibers, and a miniature 16-bit spectrophotometer [10].

In a competitive system, as shown in Fig. 3.1a, two specific DNA probes are required, too. They are labeled with donor and acceptor as the same as those in a sandwich structure. The difference from a sandwich system is that these two DNA probes are complementary to each other and one of them is complementary to the target DNA. In the presence of target DNA, it can compete with probe A to hybridize with probe B to form a more stable duplex. The donor is thus far away from acceptor producing a signal response to the target DNA. Krull demonstrated the use of red-emitting streptavidin-coated QDs (QD(605)) as donors in FRET to introduce a competitive displacement-based assay for the detection of oligonucleotides. QD–DNA bioconjugates featuring 25-mer probe sequences diagnostic of Hsp23 were prepared as capture probe. The hybridization between capture probes and dye-labeled (Alexa Fluor 647) reporter sequences which were partly complementary to capture probe provided a FRET-sensitized emission signal due to proximity of the QD and dye. After competition with target sequence, fully complementary recognition motif embedded within a 98-mer displacer sequence, a nM level for competitive displacement hybridization assays for in vitro DNA analysis was achieved [11].

Molecular beacons are oligonucleotide hybridization probes that can report the presence of specific nucleic acids in homogenous solutions. They are hairpin-shaped molecules with a fluorescent label and quencher at each end. QDs are usually employed in MBs as the donor for nucleic acid detection. As shown in Fig. 3.1c, the probe sequence is complementary to target DNA sequence. In the absence of target DNA, a hairpin-like stem-loop structure is formed, causing FRET took place. Once target DNA exists, such hairpin structure is changed into a duplex by the hybridization between probe sequence and target DNA. This causes the separation of QD donor from the quencher, giving a "signal on" response to the target DNA. Mattoussi et al. described the synthesis and characterization of a thiol-reactive hexahistidine peptidic linker that could be chemically attached to thiolated-DNA MB and mediated their self-assembly to CdSe/ZnS core–shell QDs. The hairpin DNA stem structure brought the dye acceptor into close proximity r of the QD establishing efficient FRET. The presence of DNA complementary to the MB would unwind the stem-loop structure altering the donor–acceptor distance to r' and changing the FRET efficiency (Fig. 3.4a) [12]. Similarly, Chen et al. described a hybrid fluorescent nanoprobe composed of a nuclease-resistant MB backbone, CdSe/ZnS core–shell QDs as donors, and gold nanoparticles (Au NPs) as quenchers, for the real-time visualization of virus replication in living cells (Fig. 3.4b). A hexahistidine-appended Tat peptide self-assembled onto the QD surface was employed for the noninvasive delivery of the nanoprobe. Upon the existence of the target sequence, coxsackievirus B6 (CVB 6) genome, a 7.3-fold increase in fluorescent signal could be achieved and the real-time detection of infectious viruses as well as the real-time visualization of cell-to-cell virus spreading could be realized [13]. For multiple applications, QDs with different fluorescent emissions have been linked to different MBs, allowing for simultaneous detection within one solution [14]. MBs are extremely target specific, primarily because of the competition

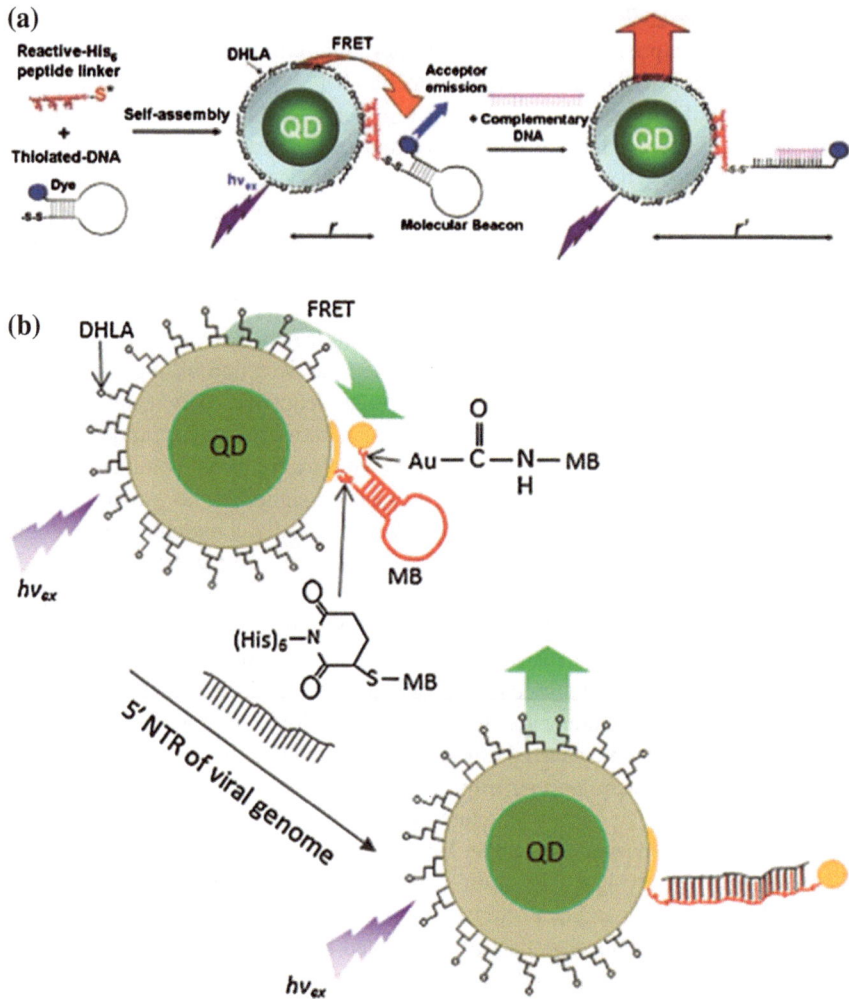

Fig. 3.4 A schematic representation of His$_6$-peptide-linker-facilitated self-assembly of a molecular beacon structure with or without the complementary DNA. Reprinted with the permission from Ref. [12]. Copyright 2007 American Chemical Society (**a**) and the QD–MB–AuNP probe with or without presence of the complementary viral RNA (**b**). Reproduced with permission from Ref. [13]. Copyright 2010, Royal Society of Chemistry

between internal hybridization within the stem structure and hybridization between the target and the loop structure [15]. Owing to such high specificity, MB has been commercially used in PCR or reverse transcriptase PCR (RT-PCR) kit. As illustrated above, fluorophore and quencher are tagged to each end of the MB. When the beacon unfolds in the presence of the complementary target sequence, the fluorescence of fluorophore will be recovered. The amount of fluorescence at any given cycle, or following cycling, depends on the amount of specific product and can be easily detected in a thermal cycler. For quantitative PCR, MBs bind to the amplified target following each cycle of amplification and the resulting signal is proportional to the amount of template. Fluorescence is monitored and reported during each annealing step when the beacon is bound to its complementary target. This information is then used during PCR or RT-PCR experiments to quantify initial copy number. Although lots of QD-based MB systems for DNA, protein, and tumor cell detection have been well documented, no commercial kit has been found in the market yet.

A phenomenon that electron transfer between QDs and DNA induces QD emission quenching is attractive for the development of label-free DNA detection and delivery systems. Clapp et al. designed a pH-responsive pentablock copolymer to form stable complexes with plasmid DNA via tertiary amine segments for potential use in gene delivery monitoring. As shown in Fig. 3.5, QDs can be quenched by the free pentablock copolymer or free DNA, but not by penta/DNA complex. Once polyplex dissociates, the released pentablock copolymer and DNA will lead to QD quenching, and thus, polyplex dissociation can be monitored with the decrease in QD fluorescence. Good behavior of QDs in monitoring the dissociation of pentablock copolymer/DNA polyplexes in vitro was demonstrated in this report, but further application for studying the release of DNA within cells did not realize [16]. Weil et al. prepared a protein-derived biopolymer coating for

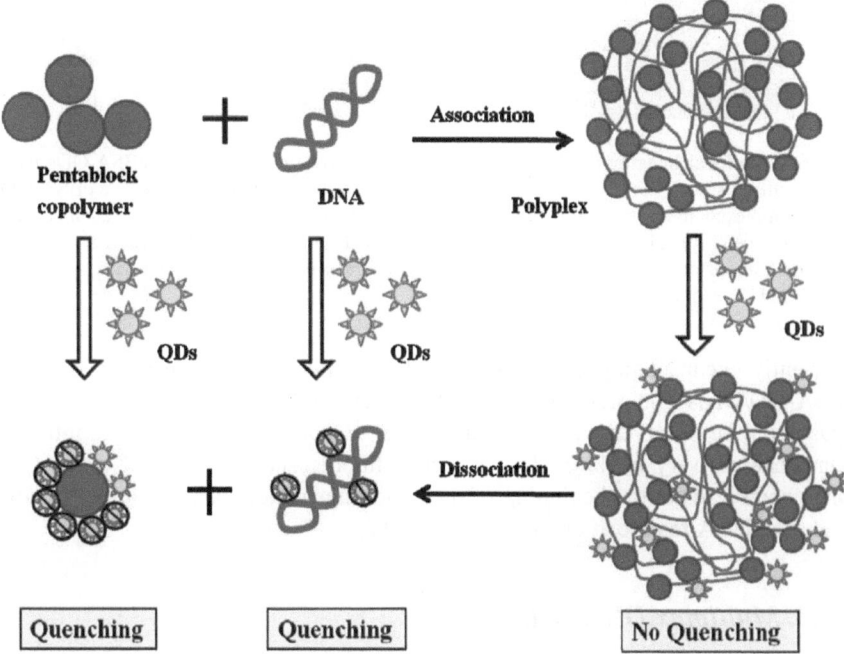

Fig. 3.5 Schematic illustration of the mechanism of sensing pentablock copolymer/DNA polyplex dissociation using QDs. QDs can be quenched by the free pentablock copolymer and/or free DNA, but not by penta/DNA polyplex. Once polyplex dissociates, the released pentablock copolymer and DNA will lead to QD quenching in such a way that polyplex dissociation can be monitored with the decrease in QD fluorescence. Reproduced with permission from Ref. [16]. Copyright 2011, American Chemical Society

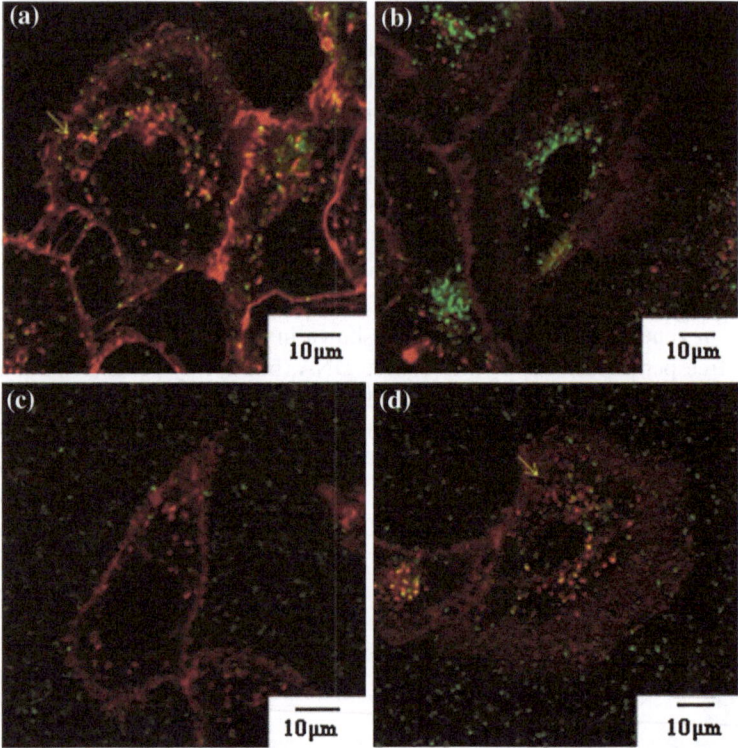

Fig. 3.6 Confocal imaging of cBSA-QDs and cBSA-QD/DNA complexes. **a** cBSA-QDs incubated with A549 cells for 5 h. **b** cBSA-QDs incubated with A549 cells for 24 h. **c** cBSA-QD/DNA complexed incubated with A549 cells for 5 h. **d** cBSA-QD/DNA complexed incubated with A549 cells for 24 h. Cell membranes were labeled with CellMask Deep Red tracker and are shown in *red color*, and cBSA-QDs are shown as *green points*. Reproduced with permission from Ref. [17]. Copyright 2012, Wiley

efficiently encapsulating QDs which are highly photoresponsive to DNA. Once cBSA-QD/DNA complex were formed, a decrease in the emission intensity of QDs could be observed with increasing DNA content. They realized the intracellular gene delivery tracking. The quenched fluorescence intensity of cBSA-QD/DNA complexes at the beginning of the gene transfection experiment was recovered after DNA release after 24 h (Fig. 3.6) [17].

3.1.2 Multiplex DNA Detection

A powerful advantage of QDs in DNA sensing lies in their ability in multiplex DNA detection, which is ascribed to their size-tunable photoluminescence and the broad absorption spectra with narrow emission bands. Broad absorption spectra allow multiple QDs to be excited with a single light source, simplifying instrumental design, increasing detection speed, and lowering cost. QD emission bands can be as narrow as 20 nm in the visible range, enabling distinct signals to be detected simultaneously with very little cross talk [18]. In 2001, Nie's group achieved multicolor optical coding for biological assays by embedding different-sized QDs into polymeric microbeads at precisely controlled ratios. They designed a model DNA hybridization system using oligonucleotide probes and triple-color-encoded beads and the coding signals could identify different DNA sequences [19].

In 2005, Zhang et al. developed a sandwich type DNA nanosensor based on single quantum dot. They chose CdSe/ZnS core–shell nanocrystals as donors and Cy5 as acceptors. As shown in Fig. 3.7, reporter probe was labeled with Cy5, and capture probe was modified with biotin to conjugate with streptavidin functionalized QDs. When a target DNA was present in solution, it was sandwiched by the two probes. Several sandwiched hybrids were then captured by a single QD through biotin–streptavidin binding, resulting in a local concentration of targets in a nanoscale domain. The occurrence of FRET enabled the detection of low concentrations of DNA in a separation-free format. The functions of QD are not only as a FRET energy donor but also a target concentrator to amplify the target signal

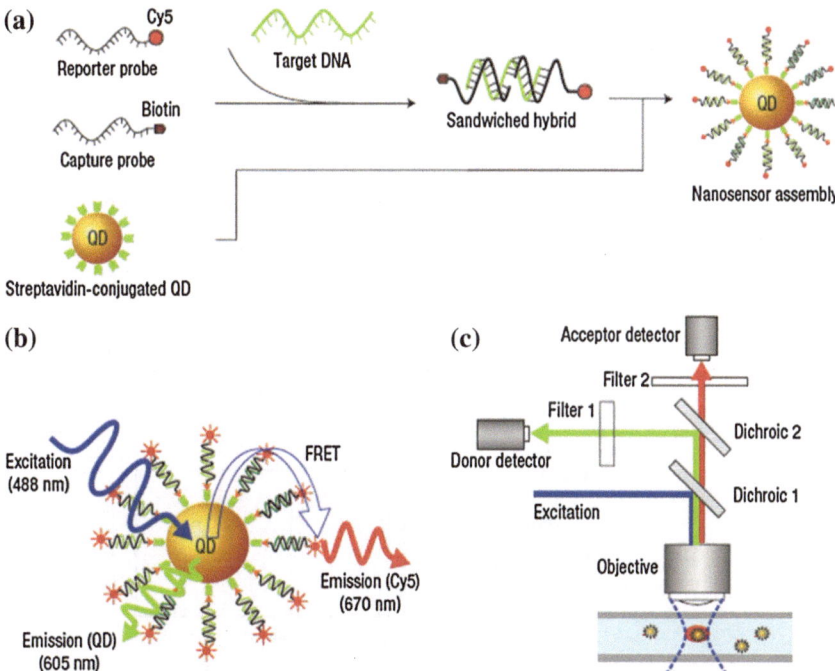

Fig. 3.7 Schematic of single QD-based DNA nanosensors. **a** Conceptual scheme showing the formation of a nanosensor assembly in the presence of targets. **b** Fluorescence emission from Cy5 on illumination on QD caused by FRET between Cy5 acceptors and a QD donor in a nanosensor assembly. **c** Experimental setup. Reproduced with permission from Ref. [21]. Copyright 2005, Macmillan Publishers Ltd

[20]. The size-tunable photoluminescence and the broad absorption spectra of QDs enable their great promise in multiplex DNA biosensing. In 2010, they successfully applied this similar proposal for multiplex DNA detection [21].

Recently, they further improved their design and described a liposome/QD-based single-particle detection technique for multiple DNA targets with attomolar sensitivity even without the involvement of any amplification step. As shown in Fig. 3.8, the carboxyl-functionalized liposome/QD (L/QD) complex and carboxyl-modified magnetic beads were covalently conjugated with the amino-terminated oligonucleotides, producing the reporter probe and the capture probe, respectively. The presence of target DNA leads to the generation of a sandwich hybrid containing L/QD complexes and results in the release of QDs, which can be sensitively counted by single-particle detection. They designed two sets of probes for the simultaneously determination of HIV-1 and HIV-2 [22].

QD barcodes are another popular approach for multiplexed DNA sensing. Chan et al. used the continuous flow focusing technique to create over 100 different

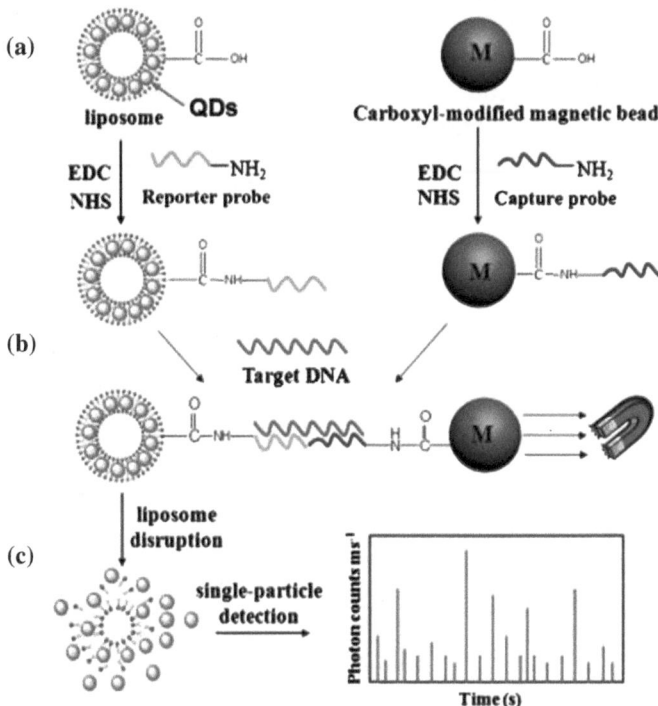

Fig. 3.8 Design principle of liposome/QD complex-based single-particle detection technique. **a** L/QD complexes, L/QD complex-tagged reporter probes and magnetic bead-modified capture probes were prepared first. **b** Formation of sandwich hybrids in the presence of target DNA and further purification by magnet. **c** Release of QDs from L/QD complex and subsequent measurement by single-particle detection. Reproduced with permission from Ref. [22]. Copyright 2013, American Chemical Society

barcodes using combinations of different emitting QDs. For application, they developed QD barcode-based assay for multiplex analysis of nine different gene fragments from pathogens such as hepatitis B (HBV), SK102 HIV-1, and HCV. Nine barcodes were prepared by mixing a combination of two different emitting QDs (500 and 600 nm) with the polymer poly(styrene-co-maleic anhydride) in chloroform. Different emitting QD barcodes were conjugated with different capture strands, and a secondary oligonucleotide was conjugated with the dye Alexa Fluor 647. A library of QD barcodes conjugated with capture strands are mixed with the secondary oligonucleotide–Alexa Fluor 647 (denoted as SA). When the target sequence was introduced, a sandwich structure of QD barcode-capture strand/target sequence/SA was formed. By measuring the optical emission of this assembled complex in a flow cytometer, a signal response to the target sequence was observed. It showed the great potential of QD in rapid gene mapping and infections disease detection [23].

3.2 QDs for RNA Detection

Ribonucleic acid is a ubiquitous family of large biological molecules that perform multiple vital roles in the coding, decoding, regulation, and expression of genes, including mRNA, tRNA, rRNA, snRNAs, and other noncoding RNAs. Together with DNA, RNA comprises the nucleic acids, which, along with proteins, constitute the three major macromolecules essential for all known forms of life. Like DNA, RNA is assembled as a chain of nucleotides. One of the major differences between DNA and RNA is the sugar, with 2-deoxyribose being replaced by the alternative pentose sugar ribose in RNA. The four bases found in DNA are adenine (abbreviated A), cytosine (C), guanine (G), and thymine (T). A fifth pyrimidine base, called uracil (U), usually takes the place of thymine in RNA and differs from thymine by lacking a methyl group on its ring. Table 3.1 shows the other differences between DNA and RNA.

Some noncoding RNAs such as siRNA, miRNA, and mRNA have attracted more attention because these RNAs play important roles in regulating proteins and associated with various types of human cancers [24–26]. The sensitive and selective detection of RNAs is of great importance in the early clinical diagnosis of cancers, as well as drug discovery. Northern blotting, quantitative, real-time PCR (qRT-PCR), and microarray-based hybridization are the widely used standard methods for analyzing RNAs [27–31]. However, these methods have some limitations such as poor reproducibility with interference from cross-hybridization, low selectivity, insufficient sensitivity, time-consuming, or large amounts of sample required. Thus, the innovative new tools for rapid, specific, and sensitive detection of RNAs are an important field of research. As we know, although some differences exist between DNA and RNA, the chemical structure of RNA is very similar to that of DNA. Therefore, we can detect RNA according to the methods of DNAs. QDs have been successfully conjugated with DNA and used in many applications [32, 33]. The

Table 3.1 Comparison between DNA and RNA

	DNA	RNA
Stands for	Deoxyribonucleic acid	Ribonucleic acid
Definition	A nucleic acid that contains the genetic instructions used in the development and functioning of all modern living organisms (scientists believe that RNA may have been the main genetic material in primitive life forms)	A single-stranded chain of alternating phosphate and ribose units with the bases adenine, guanine, cytosine, and uracil bonded to the ribose. RNA molecules are involved in protein synthesis and sometimes in the transmission of genetic information
Job/role	Medium of long-term storage and transmission of genetic information	Transfer the genetic code needed for the creation of proteins from the nucleus to the ribosome
Unique features	The helix geometry of DNA is of B-form. DNA is completely protected by the body, i.e., the body destroys enzymes that cleave DNA. DNA can be damaged by exposure to ultraviolet rays	The helix geometry of RNA is of A-form. RNA strands are continually made, broken down, and reused. RNA is more resistant to damage by ultraviolet rays
Predominant structure	Double-stranded molecule with a long chain of nucleotides	A single-stranded molecule in most of its biological roles and has a shorter chain of nucleotides
Bases and sugars	Deoxyribose sugar; phosphate backbone; four bases: adenine, guanine, cytosine, and thymine	Ribose sugar; phosphate backbone. Four bases: adenine, guanine, cytosine, and uracil
Pairing of bases	A–T (adenine–thymine), G–C (guanine–cytosine)	A–U (adenine–uracil), G–C (guanine–cytosine)
Stability	Deoxyribose sugar in DNA is less reactive because of C–H bonds. Stable in alkaline conditions. DNA has smaller grooves, which makes it harder for enzymes to "attack" DNA	Ribose sugar is more reactive because of C–OH (hydroxyl) bonds. Not stable in alkaline conditions. RNA has larger grooves, which makes it easier to be attacked by enzymes
Propagation	DNA is self-replicating	RNA is synthesized from DNA when needed

unique luminescence properties of semiconductor QDs also show a great potential to develop RNA sensors by implementing the QDs as luminescent labels.

3.2.1 Direct Fluorescence Labeling

QDs have high extinction coefficient and high quantum yield, which should dramatically increase the sensitivity in theory. Therefore, it was thought that the direct labeling of miRNA with QDs could be well used in miRNA detection and applied in microarray. Liang et al. realized microRNA detection in a microarray configuration based on the hybridization of target RNA with a capture probe attached to a solid support (shown in Fig. 3.9) [34]. 3′-termini biotinylated miRNA targets hybridized with the corresponding complementary DNA probes, which were immobilized on glass slides. Streptavidin-modified QDs were conjugated with the biotin–miRNA to read out signal (Fig. 3.9a and b). Analysis of a model system indicated that the detection limit for analyzing miRNA was about 0.4 fmol and the detection dynamic range spanned across two orders of magnitude, from 156 to 20,000 pM (Fig. 3.9c). Moreover, the method was applied to develop an assay for profiling 11 miRNA targets from rice (Fig. 3.9d). However, the lack of orthogonal conjugation methods for attaching miRNAs to QDs did not allow the use of different QD colors.

To achieve signal enhancement and multiplexed analysis, QD nanobarcode-based microbead random array platform for accurate and reproducible gene expression profiling in a high-throughput and multiplexed format was developed as shown in Fig. 3.10 [35]. Four different sizes (and thus four different fluorescent colors) of Qdots, with emissions at 525, 545, 565, and 585 nm, are mixed with a polymer

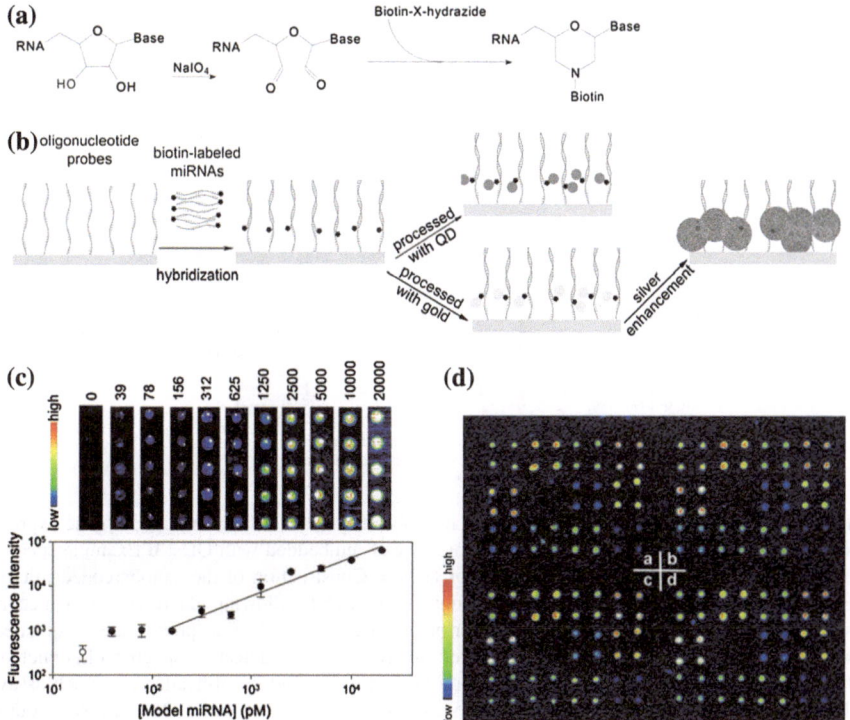

Fig. 3.9 Analysis of miRNA by means of fluorescent QDs. **a** Principle of labeling miRNA at 3′ termini with biotin. **b** Organization of the streptavidin-labeled QDs on a DNA/miRNA duplex bound to a glass support. **c** Fluorescence intensities detected upon analyzing different concentrations of a target miRNA (*upper panel*) and derived calibration curve (*lower panel*). **d** Analysis in an array format of 11 target miRNAs from rice. Reproduced with permission from Ref. [34]. Copyright 2005, Oxford University Press

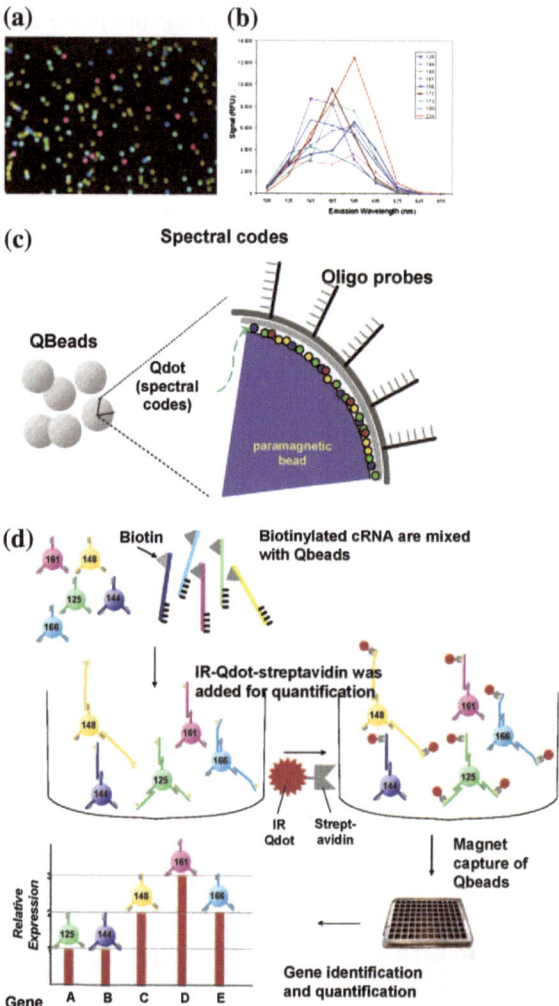

Fig. 3.10 Schematics of QD-nanobarcoded microbead system for high-throughput gene expression analysis. **a** Pseudocolor picture of the microbeads embedded with QDs. **b** Example spectra of the beads coded with different mixture of QDs. **c** Construction of the nanobarcoded microbeads. Each bead has a distinctive ratio of four different QDs, allowing identification by a characteristic spectral nanobarcode. The transcript-specific oligonucleotide probes are conjugated with the bead surface. Therefore, each spectral-barcoded bead detects a specific oligonucleotide determined by the probe. **d** Gene expression monitoring and quantification sandwich assay. The nanobarcoded microbead-attached oligo probes capture biotinylated cRNA sample through hybridization, the cRNA is further sandwiched by the 655-nm streptavidin QDs (or 705, 800 nm) to be quantified. The gene expression variation is measured by fluorescence levels upon imaging with the Mosaic scanner (Quantum Dot Corp.). Reproduced with permission from Ref. [35]. Copyright 2006, American Chemical Society

and coated onto the microbeads to generate a nanobarcoded microbead termed as QBeads. Gene-specific oligonucleotide probes are conjugated with the surface of each spectrally nanobarcoded bead to create a multiplexed panel, and biotinylated cRNAs are generated from sample total RNA and hybridized to the gene probes on the microbeads. A fifth streptavidin Qdot (655 nm or infrared Qdot) binds to biotin on the cRNA, acting as a quantification reporter. Target identity was decoded based on spectral profile and intensity ratios of the four coding Qdots (525, 545, 565, and 585 nm). The intensity of the 655 nm Qdot reflects the level of biotinylated cRNA captured on the beads and provides the quantification for the corresponding target gene. It provides increased flexibility, convenience, and cost-effectiveness in comparison with conventional gene expression profiling methods.

3.2.2 Foster (or Fluorescence) Resonance Energy Transfer System

Several studies have demonstrated the effective use of QD FRET donors to detect small analytes by utilizing a common strategy that relies on conjugating QDs with target-binding receptors, which can be either proteins [36, 37], antibody fragments [38, 39], or DNA aptamers [40, 41]. FRET and QDs were also employed for RNA analysis. For example, a single-stranded siRNA conjugated with QD was designed and used as a hybridization probe for the development of a comparatively simple and rapid procedure for preliminary screening of highly effective siRNA sequences for RNA interference (RNAi) in mammalian cells (Fig. 3.11) [42]. The target mRNA was amplified in the presence of Cy5-labeled nucleotides, and Cy5-labeled mRNA served as a hybridization sample. The accessibility and affinity of the siRNA sequence for the target mRNA site were determined by FRET between a QD (donor) and a fluorescent dye molecule (Cy5, acceptor) localized at an appropriate distance from each other when hybridization occurred. The FRET signal was observed only when there was high accessibility between an antisense siRNA and a sense mRNA and did not appear in the case of mismatch siRNAs. This method can markedly facilitate the screening of truly effective siRNAs and significantly shorten the time-consuming siRNA screening procedures.

To improve the detection sensitivity, flexibility, and adaptability, various new strategies have been developed, such as rolling circle amplification [43], isothermal amplification [44], and isothermal strand-displacement polymerase reaction [45]. A novel miRNA detection method based on the two-stage exponential amplification reaction (EXPAR) and a single QD-based nanosensor was developed (Fig. 3.12) [46]. EXPAR provides high amplification efficiency, which can rapidly amplify short oligonucleotides (10^6–10^9-fold) within minutes. The two-stage EXPAR involved two templates and two-stage amplification reactions under isothermal conditions [47]. The first-stage reaction (Fig. 3.12a, b) was an exponential amplification with the involvement of the X'–X' template, which enabled the amplification of miRNA. The second-stage reaction (Fig. 3.12c) was a linear amplification with the

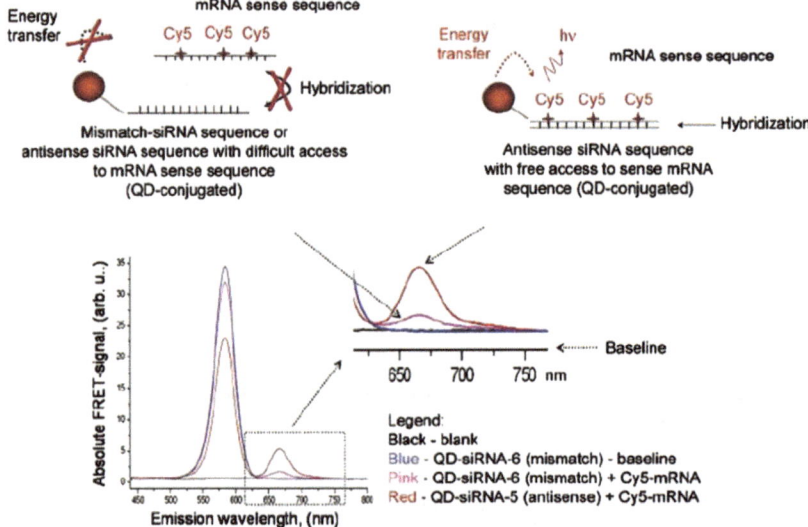

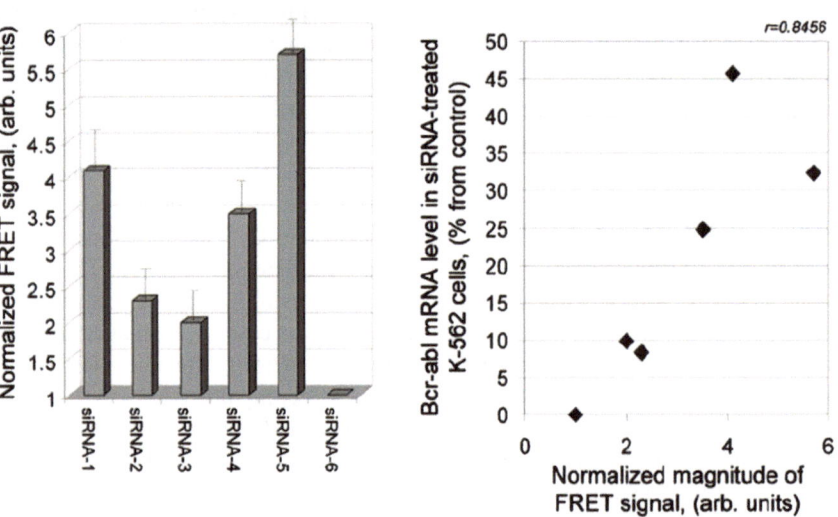

Fig. 3.11 Efficiency of interaction between siRNA hybridization probes and bcr/abl mRNA hybridization samples-FRET analysis. **a** Typical fluorescent spectra of nonhybridized and hybridized QD-labeled siRNA probe and Cy5-labeled mRNA sample (λex) 420 nm. 1: QD-siRNA-6 (mismatch). 2: QD-siRNA-6 plus Cy5-labeled mRNA. 3: QD-siRNA-5 plus Cy5-labeled mRNA. **b** FRET between QD-labeled siRNAs and Cy5-labeled target mRNA. The results on histograms represent the mean (SD from seven-independent experiments for siRNA-1, 2, and 6, and from six independent experiments for siRNA-3, 4, and 5. **c** Correlation between FRET data (obtained in this study) and effect of anti-bcr/abl siRNAs on the level of target mRNA (obtained from the literature), r is the correlation coefficient. Reproduced with permission from Ref. [42]. Copyright 2005, American Chemical Society

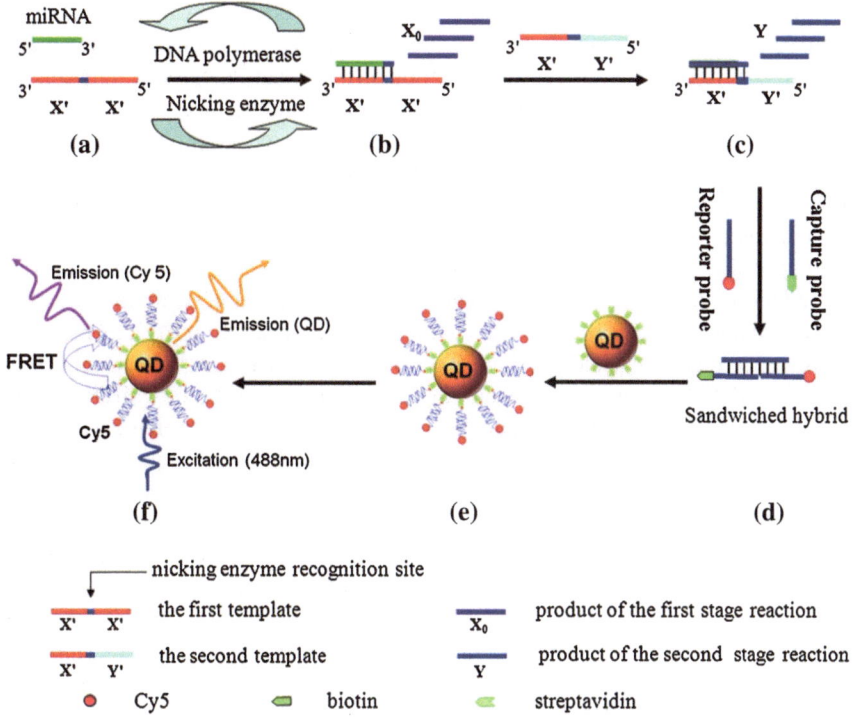

Fig. 3.12 Scheme of the miRNA assay based on the two-stage EXPAR and single QD-based nanosensor. Reproduced with permission from Ref. [46]. Copyright 2012, American Chemical Society

involvement of the X′–Y′ template, which enabled the conversion of miRNA to the reporter oligonucleotide Y. The newly formed reporter oligonucleotide Y cannot in turn prime the X′–Y′ template, thus resulting in a linear amplification (Fig. 3.12c). After amplification, the reporter oligonucleotide Y was sandwiched by a biotinylated capture probe and a Cy5-labeled reporter probe (Fig. 3.12d). This sandwich hybrid was then assembled on the surface of a 605QD to form the 605QD/reporter oligonucleotide Y/Cy5 complex through specific biotin–streptavidin binding (Fig. 3.12e). When this complex was excited by a 488-nm argon laser, the fluorescence signals of 605QD and Cy5 were observed simultaneously due to FRET from 605QD to Cy5 (Fig. 3.12f). Isothermal amplification and the single QD-based nanosensor offer improved sensitivity and selectivity for miRNA assay.

3.2.3 Sensing Based on DNA-Scaffolded Metal Nanoclusters

Recently, the emergence of noble metal nanoclusters as a novel type of robust and promising fluorescence materials offers great potential for biological labeling, biosensing, bioimaging, and diagnostic applications [48–50]. They typically consist

of a few atoms and possess molecular-like optical properties, which are complementary to those of organic dyes and QDs. Particularly, oligonucleotide-templated
nanoclusters have drawn increasing attention as a fascinating class of fluorophores
due to their amazing features of facile synthesis, ultrafine size, and outstanding
spectral and photophysical properties. Using the fluorescence properties of DNA–
nanosilver clusters (DNA–AgNC), a DNA–AgNC probe that can detect the presence of target miRNA has been designed (Fig. 3.13) [51]. On the basis of the work
of Richards et al., a sequence was chosen as scaffold to successfully create a red-
emitting AgNC (DNA-12nt-RED: 5′-CCTCCTTCCTCC-3′) [52]. DNA-12nt-RED
created red-emissive AgNCs, with an emission maximum at 620 nm, molar extinction coefficient of 120,000 $M^{-1}cm^{-1}$, fluorescence decay time of 2.23 ns, and a
fluorescence quantum yield of 32 %. Therefore, the DNA-12nt-RED would be an
ideal candidate to be used for developing a DNA-based probe for miRNA detection. In order to create the probe for detecting miRNA, the complementary DNA
sequence of RNA-miR160 was attached to the DNA-12nt-RED sequence. Just like
DNA-12nt-RED, after addition of $AgNO_3$ and reduction with $NaBH_4$, the DNA-
12nt-RED-160 probe displayed strong red emission from the AgNCs. Moreover,
the speed of AgNC formation is faster and the overall fluorescence intensity is
about 100 times higher in the DNA-12nt-RED-160 probe versus DNA-12nt-RED.
The reason is likely that DNA-12nt-RED-160 probe could form a self-dimer or a
hairpin structure in the absence of miRNA. In the presence of an increasing concentration of RNA-miR160 sequences, the observed red fluorescence decreased,
demonstrating that the DNA-12nt-RED-160 probe can be used for detecting the
RNA-miR160 target molecules by monitoring the generated red fluorescence of
the AgNCs.

A facile detection of microRNA by target-assisted isothermal exponential amplification (TAIEA) coupled with fluorescent DNA-scaffolded AgNC was also investigated (Fig. 3.14) [53]. The TAIEA reaction utilizes a unimolecular DNA containing
three functional domains as the amplification template, polymerases, and nicking
enzymes as mechanical activators and target miRNA as the trigger, which enables

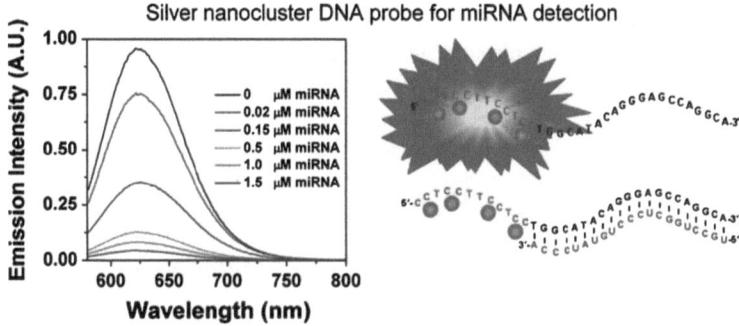

Fig. 3.13 Scheme of the miRNA assay based on silver nanocluster DNA probe. Reproduced
with permission from Ref. [51]. Copyright 2011, American Chemical Society

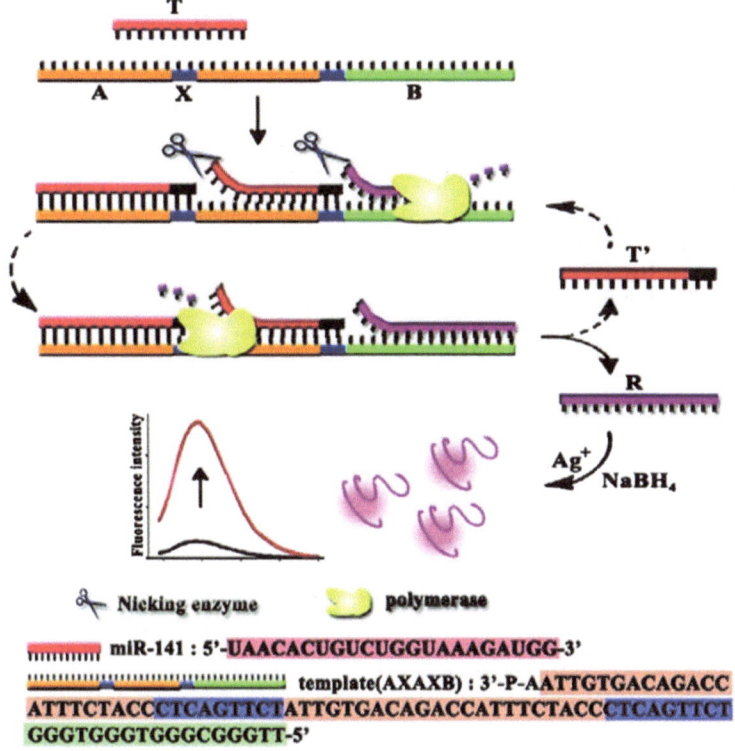

Fig. 3.14 Detection of miRNA with attomolar sensitivity based on target-assisted isothermal exponential amplification (TAIEA) coupled with fluorescent DNA-templated AgNC probe. Reproduced with permission from Ref. [53]. Copyright 2012, American Chemical Society

the conversion of miRNA to a great amount of the reporter oligonucleotides R within minutes. After amplification, the reporter oligonucleotide R was acting as a scaffold for the synthesis of fluorescent silver nanoclusters in the presence of Ag^+ through the reduction of $NaBH_4$. The DNA–AgNCs displayed fluorescence emission at 644 nm upon the excitation at 574 nm, functioning as signal indicators in a label-free and environmental-friendly format. The method reveals superior sensitivity with a detection limit of miRNA of 2 aM synthetic spike-in target miRNA under pure conditions (approximately 15 copies of a miRNA molecule in a volume of 10 μL) and can detect at least a 10 aM spike-in target miRNA in cell lysates.

DNA–AgNCs have attracted an explosion of interest in biological analysis. However, they have one major problem in practical application, that is, Ag^+ ions can easily form an insoluble product with many general anions, such as Cl^- or SO_4^{2-}, to terminate the formation of AgNCs. In addition, the DNA–AgNCs formation by the reduction of Ag^+ ions with $NaBH_4$ usually takes one to several hours. In contrast to DNA–AgNCs, the newly emerging CuNCs, selectively formed on a DNA duplex [54–56], offer excellent potential for "on the spot"

Fig. 3.15 Schematic illustration of miRNA detection using a novel fluorescence probe of dsDNA-templated copper nanoclusters as the signal output, via the target-triggered isothermal exponential amplification reaction. Reproduced with permission from Ref. [57]. Copyright 2013, Royal Society of Chemistry

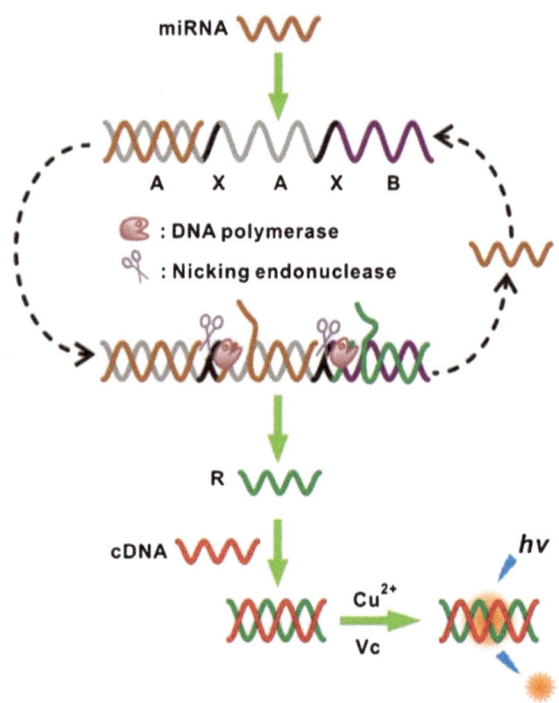

testing with a rapid and simple "mix-and-measure" format. dsDNA–CuNCs can be facilely prepared by reducing Cu^{2+} ions with ascorbic acid in the presence of a DNA duplex within fifteen minutes. More importantly, the Cu^{2+} ions are soluble in many detection environments and so have no precipitation phenomena like the Ag^{+} ions. Therefore, a facile label-free method for sensitive and selective detection of miRNAs using dsDNA–CuNCs as fluorescent reporters was developed (Fig. 3.15) [57]. In order to improve the analytical sensitivity, they have introduced the target-triggered isothermal exponential amplification reaction (TIEAR) to the proposed method. This amplification strategy has shown great potential as the point-of-care testing with high amplification efficiency under a constant temperature.

3.2.4 Sensing Based on Fluorescence In Situ Hybridization

In situ detection of RNAs is becoming increasingly important for analysis of gene expression within and between intact cells in tissues. Fluorescence in situ hybridization (FISH) of nucleic acid-labeled probes provides a direct visualization of the spatial location of specific DNA or RNA sequences at a particular cellular or chromosomal site and in tissue sections [58]. Streptavidin-labeled/biotinylated

DNA probes have been used for FISH detection of several mRNA targets [59, 60]. However, QDs containing cadmium such as CdSe or CdTe are always toxic due to the release of Cd^{2+}. In view of the inherent toxicity of the cadmium, an additional nontoxic ZnS shell was coated onto the CdSeTe/CdS QDs to minimize the toxicity of NIR-emitting QDs [35, 61]. Trioctylphosphine oxide (TOPO)-capped CdSe/ZnS QDs was prepared to reduce cytotoxicity of QDs [62]. Amine-modified oligonucleotide probes are designed and covalently attached to the carboxyl-terminated polymer-coated QDs via EDC chemistry (shown in Fig. 3.16). The resulting QD–DNA conjugates show sequence-specific hybridization with target mRNAs. QD–DNA probes exhibit excellent sensitivity to detect the low-expressing dorsal-related immunity factor gene. Importantly, multiplex FISH of ribosomal protein 49 and actin 5C using green and red QD–DNA conjugates allows the observation of cellular distribution of the two independent genes simultaneously.

3.3 QDs for DNA Microarrays

DNA microarrays (also commonly called gene chips, DNA chips, or biochips) are a collection of microscopic DNA spots attached to a solid surface, such as glass, plastic, or silicon supports [63, 64]. They were born for the simultaneous analysis of the expression levels of numerous genes in a single experiment [65]. Since their development in the mid-1990s, DNA microarrays have displayed enormous potentials in various applications, such as cancer diagnosis or drug influence on the gene expression level. DNA microarray technology has revealed a great deal about the genetic factors involved in a number of diseases, including multiple forms of cancer. At the beginning, microarrays were just employed for the identification of the differences in gene expression between normal cells and their cancerous counterparts. Later on, researchers began to apply this technology to distinguish specific subtypes of certain cancers, as well as to determine which treatment methods would most likely be effective for particular patients. This also reflected the effect of medicines which was benefit for drug selection [66, 67]. DNA microarrays are usually categorized as complementary DNA (cDNA) arrays, using either short (25–30 mer) or long ODN (60–70 mer) probes. The core principle behind microarrays lies in their ability to provide a powerful high-throughput system that allows for large-scale analysis of gene expression, genetic alterations, and signal transduction pathways, which can give important information for disease diagnosis, prognostics, and therapeutics. A typical DNA microarray usually consists of the following step: First robotically printing oligonucleotides or cDNA clone inserts onto a glass slide, then one or more fluorescent-labeled cDNA probes generated from samples are hybridized to the surface, later a laser is employed to excite the dye labels and record the fluorescent intensities by a laser confocal fluorescent scanner. Finally, the ratio of fluorescent intensities provides the basis for further meta-analyses. The feasibility

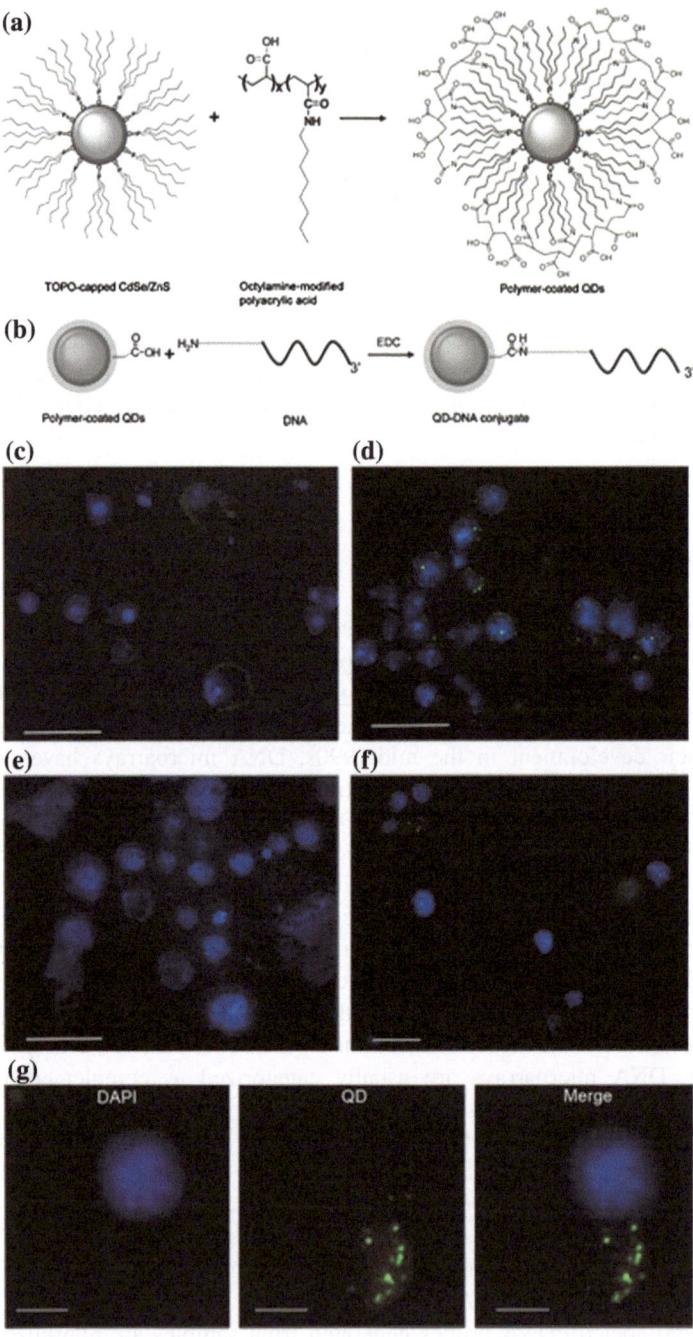

Fig. 3.16 Scheme for polymer coating of QD–TOPO (**a**) and the conjugation reaction with amine-terminated oligonucleotides using EDC-coupling reagent (**b**). Specificity of QD–DNA probes hybridization in situ to detect mRNA (Rp49) in fixed Drosophila S2 cells. **c** Control (QD555 alone, no DNA attached). **d** QD555–DNA conjugate. **e** Control experiment with QD555–noncomplementary DNA conjugate. **f** RNase A treatment 10 mg mL prior to cell fixation significantly decreased the FISH signal. **g** Representative single-cell FISH image separately displaying a DAPI-stained cell nucleus and the QD–DNA conjugates (for Rp49) exclusively located in the cytoplasm, and the merged image of the DAPI signal and QD555 signal. Reproducibility was obtained from three separate experiments. The scale bar indicates 20 mm (**c–f**) and 5 mm (**g**). Reproduced with permission from Ref. [62]. Copyright 2009, Wiley

of QDs in such DNA microarray is ascribed to the narrow multicolor emissions under a single-source excitation. Meissner et al. developed a QD-embedded microsphere-based fluid DNA microarray. CdSe/ZnS QDs embedded in polystyrene microspheres were labeled with DNA oligonucleotides for target capture, and the result was recorded by a high-speed readout flow cytometer [68]. QD-based cDNA microarray was also developed for single-nucleotide polymorphism (SNP) mutation detection in the human p53 tumor suppressor gene and multiallele detections. The authors established a model that used a SNP located at amino acid residue 248, on exon 7 of p53, which is one of the most common mutation hot spots in p53 [69]. As shown in Fig. 3.17, multicolor targeting of SNP mutations in human oncogene p53 and of human hepatitis B and C viruses with different DNA–QDs were realized, which displayed the great potential of DNA–QD conjugates as efficient probes in cDNA microarrays for a ultrafast detection of a great number of viral or bacterial pathogens simultaneously. However, this model system was not extended to real samples such as blood [70].

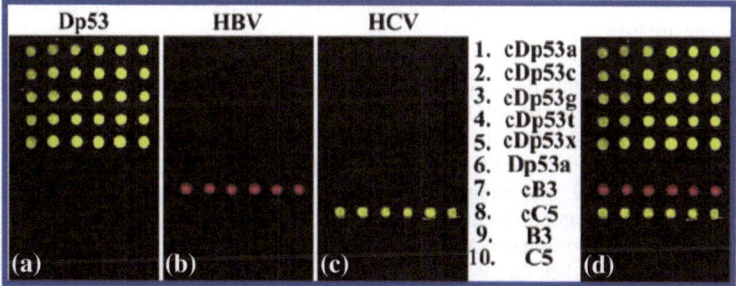

Fig. 3.17 Multicolor targeting of SNP mutations in human oncogene p53 (rows *1–6*) and of human hepatitis B and C viruses (rows *7–10*) with different DNA-QD probes. In panel **a**, *yellow* Dp53 g-QDs are targeted only toward mutations in human oncogene p53, but SNP detection is not achieved. In panel **b**, *red* B3-QDs are targeted toward the hepatitis B virus sequence. In panel **c**, *yellow* C5-QDs are targeted only toward the hepatitis C virus sequence. Panel **d** is an overlay of panels **a–c**. In both *red* and *yellow* channels, signal-to-noise ratio is >100, and no cross talk is observed. Reproduced with permission from Ref. [70]. Copyright 2003, American Chemical Society

References

1. Lu HC, Schops O, Woggon U, Niemeyer CM (2008) Self-assembled donor comprising quantum dots and fluorescent proteins for long-range fluorescence resonance energy transfer. J Am Chem Soc 130(14):4815–4827
2. Sapsford KE, Berti L, Medintz IL (2006) Materials for fluorescence resonance energy transfer analysis: beyond traditional donor-acceptor combinations. Angew Chem Int Ed Engl 45(28):4562–4589
3. Tu DT, Liu LQ, Ju Q, Liu YS, Zhu HM, Li RF, Chen XY (2011) Interfacial self-assembly of cell-like filamentous microcapsules. Angew Chem Int Ed 50(28):6306–6310
4. Jares-Erijman EA, Jovin TM (2003) FRET imaging. Nat Biotechnol 21(11):1387–1395
5. Clapp A, Medintz IL, Fisher BR, Anderson GP, Mattoussi H (2005) Can luminescent quantum dots be efficient energy acceptors with organic dye donors? J Am Chem Soc 127:1242–1250
6. Chen GW, Song FL, Xiong XQ, Peng XJ (2013) Fluorescent nanosensors based on fluorescence resonance energy transfer (FRET). Ind Eng Chem Res 52(33):11228–11245
7. Noor MO, Krull UJ (2013) Paper-based solid-phase multiplexed nucleic acid hybridization assay with tunable dynamic range using immobilized quantum dots as donors in fluorescence resonance energy transfer. Anal Chem 85(15):7502–7511
8. Zhang CL, Xu J, Zhang SM, Ji XH, He ZK (2012) One-pot synthesized DNA-CdTe quantum dots applied in a biosensor for the detection of sequence-specific oligonucleotides. Chem Eur J 18:8296–8300
9. Cui DX, Pan BF, Zhang H, Gao F, Wu RN, Wang JP, He R, Asahi T (2008) Self-assembly of quantum dots and carbon nanotubes for ultrasensitive DNA and antigen detection. Anal Chem 80:7996–8001
10. Chou CC, Huang YH (2012) Nucleic acid sandwich hybridization assay with quantum dot-induced fluorescence resonance energy transfer for pathogen detection. Sensors 12(12):16660–16672
11. Vannoy CH, Chong L, Le C, Krull UJA (2013) Competitive displacement assay with quantum dots as fluorescence resonance energy transfer donors. Anal Chim Acta 759:92–99
12. Medintz IL, Berti L, Pons T, Grimes AF, English DS, Alessandrini A, Facci P, Mattoussi H (2007) A reactive peptidic linker for self-assembling hybrid quantum dot-DNA bioconjugates. Nano Lett 7(6):1741–1748
13. Yeh H-Y, Yates MV, Mulchandani A, Chen W (2010) Molecular beacon—quantum dot—Au nanoparticle hybrid nanoprobes for visualizing virus replication in living cells. Chem Commun 46:3914–3916
14. Kim JH, Chaudhary S, Ozkan M (2007) Multicolor hybrid nanoprobes of molecular beacon conjugated quantum dots: FRET and gel electrophoresis assisted target DNA detection. Nanotechnology 18:195105–195112
15. Tan L, Li Y, Drake TJ, Moroz L, Wang K, Li J, Munteanu A, Chaoyong JY, Martinez K, Tas W (2005) Molecular beacons for bioanalytical applications. Analyst 130:1002–1005
16. Zhang BQ, Zhang YJ, Mallapragada SK, Clapp AR (2011) Sensing polymer/DNA polyplex dissociation using quantum dot fluorophores. ACS Nano 5(1):129–138
17. Wu YZ, Eisele K, Doroshenko M, Algara-Siller G, Kaiser U, Koynov K, Weil T (2012) A quantum dot photoswitch for DNA detection, gene transfection, and live-cell imaging. Small 8(22):3465–3475
18. Smith AM, Dave S, Nie SM, True L, Gao XH (2006) Multicolor quantum dots for molecular diagnostics of cancer. Expert Rev Mol Diagn 6(2):231–244
19. Han MY, Gao XH, Su JZ, Nie SM (2001) Quantum-dot-tagged microbeads for multiplexed optical coding of biomolecules. Nat Biotech 19:631–635
20. Zhang CY, Yeh HC, Kuroki MT, Wang TH (2005) Single-quantum-dot-based DNA nanosensor. Nature Mater 4:826–831
21. Zhang CY, Hu J (2010) Single quantum dot-based nanosensor for multiple DNA detection. Anal Chem 82:1921–1927

22. Zhou J, Wang QX, Zhang CY (2013) Liposome-quantum dot complexes enable multiplexed detection of attomolar DNAs without target amplification. J Am Chem Soc 135:2056–2059
23. Giri S, Sykes EA, Jennings TL, Chan WCW (2011) Rapid screening of genetic biomarkers of infectious agents using quantum dot barcodes. ACS nano 5(3):1580–1587
24. Cheng YQ, Zhang X, Li ZP, Jiao XX, Wang YC, Zhang YL (2009) Highly sensitive determination of microRNA using target-primed and branched rolling-circle amplification. Angew Chem Int Ed 121:3318–3322
25. Wang H, Ach RA, Curry B (2007) Direct and sensitive miRNA profiling from low-input total RNA. RNA 13:151–159
26. Arenz C (2006) MicroRNAs—zuknftige Wirkstoff-Targets? Angew Chem Int Ed 118:5170–5172
27. Raymond CK, Roberts BS, Garrett-Engele P, Lim LP, Johnson JM (2005) Simple, quantitative primer-extension PCR assay for direct monitoring of microRNAs and short-interfering RNAs. RNA 11:1737–1744
28. Lagos-Quintana M, Rauhut R, Lendeckel W, Tuschl T (2001) Identification of novel genes coding for small expressed RNAs. Science 294:853–858
29. Tu YQ, Wu P, Zhang H, Cai CX (2012) Fluorescence quenching of gold nanoparticles integrating with a conformation-switched hairpin oligonucleotide probe for microRNA detection. Chem Commun 48:10718–10720
30. Li W, Ruan KC (2009) MicroRNA detection by microarray. Anal Bioanal Chem 394:1117–1124
31. Gao ZQ, Peng YF (2011) MicroRNA detection by microarray. Biosens Bioelectron 26:3768–3773
32. Cui ZQ, Ren Q, Wei HP, Chen Z, Deng JY, Zhang ZP, Zhang XE (2011) Quantum dot–aptamer nanoprobes for recognizing and labeling influenza A virus particles. Nanoscale 3:2454–2457
33. Bi S, Zhou H, Zhang SS (2010) A novel synergistic enhanced chemiluminescence achieved by a multiplex nanoprobe for biological applications combined with dual-amplification of magnetic nanoparticles. Chem Sci 1:681–687
34. Liang RQ, Li W, Li Y, Tan CY, Li JX, Jin YX, Ruan KC (2005) An oligonucleotide microarray for microRNA expression analysis based on labeling RNA with quantum dot and nano-gold probe. Nucleic Acids Res 33:e17
35. Eastman PS, Ruan WM, Doctolero M, Nuttall R, Feo GD, Park JS, Chu JSF, Cooke P, Gray JW, Li S, Chen FQF (2006) Qdot nanobarcodes for multiplexed gene expression analysis. Nano Lett 6:1059–1064
36. Li LL, Chen Y, Lu Q, Ji J, Shen YY, Xu M, Fei R, Yang GH, Zhang K, Zhang JR, Zhu JJ (2013) Electrochemiluminescence energy transfer-promoted ultrasensitive immunoassay using near-infrared-emitting CdSeTe/CdS/ZnS quantum dots and gold nanorods. Scientific Reports 3:1529
37. Wagh A, Jyoti F, Mallik S, Qian S, Leclerc E, Law B (2013) Polymeric nanoparticles with sequential and multiple FRET cascade mechanisms for multicolor and multiplexed imaging. Small 9(12):2129–2139
38. Zhu D, Jiang XX, Zhao CE, Sun XL, Zhang JR, Zhu JJ (2010) Green synthesis and potential application of low-toxic Mn: ZnSe/ZnS core/shell luminescent nanocrystals. Chem Commun 46:5226–5228
39. Liang GX, Pan HC, Li Y, Jiang LP, Zhang JR, Zhu JJ (2009) Near infrared sensing based on fluorescence resonance energy transfer between Mn: CdTe quantum dots and Au nanorods. Biosens Bioelectron 24(12):3693–3697
40. Sharon E, Freeman R, Willner I (2010) CdSe/ZnS quantum dots-G-quadruplex/hemin hybrids as optical DNA sensors and aptasensors. Anal Chem 82:7073–7077
41. Dong HF, Gao WC, Yan F, Ji HX, Ju HX (2010) Fluorescence resonance energy transfer between quantum dots and graphene oxide for sensing biomolecules. Anal Chem 82:5511–5517
42. Bakalova R, Zhelev Z, Ohba H, Baba Y (2005) Quantum dot-conjugated hybridization probes for preliminary screening of siRNA sequences. J Am Chem Soc 127:11328–11335

43. Bi S, Ji B, Zhang ZP, Zhu JJ (2013) Metal ions triggered ligase activity for rolling circle amplification and its application in molecular logic gate operations. Chem Sci 4:1858–1863

44. Li CP, Li ZP, Jia HX, Yan JL (2011) One-step ultrasensitive detection of microRNAs with loop-mediated isothermal amplification (LAMP). Chem Commun 47:2595–2597

45. Dong HF, Zhang J, Ju HX, Lu HT, Wang SY, Jin S, Hao KH, Du HW, Zhang XJ (2012) Highly sensitive multiple microRNA detection based on fluorescence quenching of graphene oxide and isothermal strand-displacement polymerase reaction. Anal Chem 84(10):4587–4593

46. Zhang Y, Zhang CY (2012) Sensitive detection of microRNA with isothermal amplification and a single-quantum-dot-based nanosensor. Anal Chem 84:224–231

47. Tan E, Wong J, Nguyen D, Zhang Y, Erwin B, Van Ness LK, Baker SM, Galas DJ, Niemz A (2005) Isothermal DNA amplification coupled with DNA nanosphere-based colorimetric detection. Anal Chem 77:7984–7992

48. Liu CG, Calin GA, Meloon B, Gamliel N, Sevignani C, Ferracin M, Dumitru CD, Shimizu M, Zupo S, Dono M, Alder H, Bullrich F, Negrini M, Croce CM (2004) An oligonucleotide microchip for genome-wide microRNA profiling in human and mouse tissues. Proc Natl Acad Sci USA 101:9740–9744

49. Lim LP, Lau NC, Garrett-Engele P, Grimson A, Schelter JM, Castle J, Bartel DP, Linsley PS, Johnson JM (2005) Microarray analysis shows that some microRNAs downregulate large numbers of target mRNAs. Nature 433:769–773

50. Zhang M, Guo SM, Li YR, Zuo P, Ye BC (2012) A label-free fluorescent molecular beacon based on DNA-templated silver nanoclusters for detection of adenosine and adenosine deaminase. Chem Commun 48:5488–5490

51. Yang SW, Vosch T (2011) Rapid detection of microRNA by a silver nanocluster DNA probe. Anal Chem 83:6935–6939

52. Richards CI, Choi S, Hsiang JC, Antoku Y, Vosch T, Bongiorno A, Tzeng YL, Dickson RM (2008) Oligonucleotide-stabilized Ag nanocluster fluorophores. J Am Chem Soc 130:5038–5039

53. Liu YQ, Zhang M, Yin BC, Ye BC (2012) Attomolar ultrasensitive microRNA detection by DNA-scaffolded silver-nanocluster probe based on isothermal amplification. Anal Chem 84:5165–5169

54. Rotaru A, Dutta S, Jentzsch E, Gothelf K, Mokhir A (2010) Selective dsDNA-templated formation of copper nanoparticles in solution. Angew Chem Int Ed 49:5665–5667

55. Jia X, Li J, Han L, Ren J, Yang X, Wang E (2012) DNA-hosted copper nanoclusters for fluorescent identification of single nucleotide polymorphisms. ACS Nano 6:3311–3317

56. Chen J, Liu J, Fang Z, Zeng L (2011) Random dsDNA-templated formation of copper nanoparticles as novel fluorescence probes for label-free lead ions detection. Chem Commun 48:1057–1059

57. Wang XP, Yin BC, Ye BC (2013) A novel fluorescence probe of dsDNA-templated copper nanoclusters for quantitative detection of microRNAs. RSC Adv 3:8633–8636

58. Bentolila LA, Weiss S (2006) Single-step multicolor fluorescence in situ hybridization using semiconductor quantum dot-DNA conjugates. Cell Biochem Biophys 45:59–70

59. Chan P, Yuen T, Ruf F, Gonzalez-Maeso J, Sealfon SC (2005) Method for multiplex cellular detection of mRNAs using quantum dot fluorescent in situ hybridization. Nucl Acids Res 33(18):e161

60. Tholouli E, Hoyland JA, Di Vizio D, O'Connell F, MacDermott SA, Twomey D, Levenson R, Liu YJA, Golub TR, Loda M, Byers R (2006) Imaging of multiple mRNA targets using quantum dot based in situ hybridization and spectral deconvolution in clinical biopsies. Biochem Biophys Res Commun 348:628–636

61. Shen YY, Li LL, Lu Q, Ji J, Fei R, Zhang JR, Abdel-Halim ES, Zhu J-J (2012) Microwave-assisted synthesis of highly luminescent CdSeTe@ZnS–SiO$_2$ quantum dots and their application in the detection of Cu(II). Chem Commun 48:2222–2224

62. Choi Y, Kim HP, Hong SM, Ryu JY, Han SJ, Song R (2009) In situ visualization of gene expression using polymer-coated quantum–dot–DNA conjugates. Small 5(18):2085–2091

63. Wang J (2000) From DNA biosensors to gene chips. Nucl Acids Res 28:3011–3016

64. Hahn S, Mergenthaler S, Zimmermann B, Holzgreve W (2005) Nucleic acid based biosensors: the desires of the user. Bioelectrochemistry 67:151–154
65. Albelda SM, Shepard JRE (2000) Functional genomics and expression profiling: be there or be square. Am J Respir Cell Mol Biol 23:265–269
66. Hoopes L (2008) Genetic diagnosis: DNA microarrays and caner. Nature Education 1:1
67. Shepard JRE (2006) Polychromatic microarrays: simultaneous multicolor array hybridization of eight samples. Anal Chem 78(8):2478–2486
68. Meissner KE, Herz E, Kruzelock RP, Spillman WB (2003) Quantum dot-tagged microspheres for fluid-based DNA microarrays. Phys Stat Sol (C) 4:1355–1359
69. Pfeifer GP (2000) p53 mutational spectra and the role of methylated CpG sequence. Mutat Res 450:155–166
70. Gerion D, Chen FQ, Kannan B, Fu AH, Parak WJ, Chen DJ, Majumdar A, Alivisatos AP (2003) Room-temperature single-nucleotide polymorphism and multiallele DNA detection using fluorescent nanocrystals and microarrays. Anal Chem 75(18):4766–4772

Chapter 4
Quantum Dot-Electrochemiluminescence-Based Biosensing

Abstract As newly developed inorganic materials, quantum dots (QDs) have received considerable attention because of their unique nanorelated properties including high quantum yield, simultaneous excitation with multiple fluorescence colors, and electrochemical properties. This chapter presents a general description of the electrochemiluminescence (ECL) related to QDs and their analytical application. QDs including Si nanopaticles, semiconductor nanocrystals (NCs) and recent emerged novel QDs such as graphene QDs were discussed about their ECL behaviors and mechanisms. By utilization of this excellent property, new developments and improvements of its application in DNA-based analysis are discussed. Different types of QDs with different strategies for the DNA-biosensing constructions were expatiated and compared in detail.

Keywords Quantum dots • Electrochemiluminescence • Aptamer • Biosensor

In 2002, Bard et al. found that quantum dots (QDs) could generate light emission during potential cycling or pulsing, which was called electrogenerated chemiluminescence (also known as electrochemiluminescence and abbreviated ECL) [1, 2]. ECL involves the generation of species at electrode surfaces that undergo electron-transfer reactions to form excited states, and light-producing procedure when the excited molecule decays to the ground state. It is the combination of chemiluminescence (CL) and electrochemistry. It not only holds the advantages of sensitivity and wide dynamic range inherent to conventional CL, but exhibits the advantages of the electrochemical method such as the simplicity, facility. Since the first detailed ECL investigations described by Hercules and Bard et al. in the mid-1960s [3–5], ECL has now become a powerful analytical technique and has been widely used in the areas of immunoassay [6–10], food and water testing [11, 12], biowarfare agent detection, [13, 14] DNA and aptamer biosensor [15–18]. In this chapter, the basic QDs ECL mechanisms and its application in DNA-based detection are discussed in detail.

J.-J. Zhu et al., *Quantum Dots for DNA Biosensing*, SpringerBriefs in Molecular Science, DOI: 10.1007/978-3-642-44910-9_4, © The Author(s) 2013

4.1 ECL Mechanism of QDs

ECL involves the emission of light by species that can undergo highly energetic electron-transfer reactions which are controlled by changing an electrode potential, as was schematically shown in Fig. 4.1.

In general, there are three main types of ECL: annihilation ECL, coreactant ECL, and cathodic luminescence. Annihilation ECL is the first fully studied ECL which involved electron-transfer reactions between an oxidized and a reduced species. After the emitter is electrochemically oxidized and reduced, the newly formed radical cation and anion are annihilated to form the excited-state species that emits light [3, 5]. Cathodic luminescence results from the injection of hot electrons into the aqueous electrolyte solution with the possible formation of hydrated electrons. For example, emission from Dy(III), Sm(III), and Tb(III) has been observed at oxide-covered aluminum electrodes during the reduction in hydrogen peroxide, persulfate, or oxylate in aqueous solution [19–21].

Till now, the coreactant ECL has widely been studied and used in ECL technique, and most of the commercially available ECL analytical instruments are based on coreactant ECL. Unlike the annihilation ECL, coreactant ECL is usually generated with the reaction between the luminophore species and an additionally added assistant reagent (which is called coreactant) in a single potential step or a directional potential scanning. Depending on the polarity of the applied potential, the corresponding ECL reactions could be classified as "oxidative reduction" ECL and "reductive oxidation" ECL, respectively. Compared to the annihilation ECL, the use of coreactant can enhance the ECL efficiency. The oxalate ion ($C_2O_4^{2-}$) is the first-discovered coreactant [23] and the ECL mechanism is described as follows by Rubinstein and Bard [24]:

$$Ru(bpy)_3^{2+} - e^- \rightarrow Ru(bpy)_3^{3+} \qquad (4.1)$$

Fig. 4.1 Schematic diagrams of electrogenerated chemiluminescence. Reprinted with permission from Ref. [22]. Copyright 2007 Elsevier

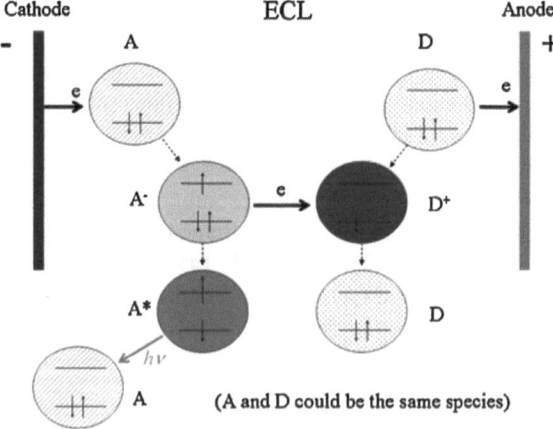

$$Ru(bpy)_3^{3+} + C_2O_4^{2-} \rightarrow Ru(bpy)_3^{2+} + C_2O_4^{\cdot-} \qquad (4.2)$$

$$C_2O_4^{\cdot-} \rightarrow CO_2^{\cdot-} + CO_2 \qquad (4.3)$$

$$Ru(bpy)_3^{3+} + CO_2^{\cdot-} \rightarrow Ru(bpy)_3^{2+*} + CO_2 \qquad (4.4)$$

$$Ru(bpy)_3^{2+} + CO_2^{\cdot-} \rightarrow Ru(bpy)_3^{+} + CO_2 \qquad (4.5)$$

$$Ru(bpy)_3^{3+} + Ru(bpy)_3^{+} \rightarrow Ru(bpy)_3^{2+*} + Ru(bpy)_3^{2+} \qquad (4.6)$$

$$Ru(bpy)_3^{2+*} \rightarrow Ru(bpy)_3^{2+} + h\nu \qquad (4.7)$$

Other common coreactants include peroxydisulfate (persulfate, $S_2O_8^{2-}$), tri-n-propylamine (TPrA) and other amine-related derivatives, hydrogen peroxide (H_2O_2). The main organic luminants contain luminal, tris(2,2′-bipyridine) ruthenium(II) $\left(Ru(bpy)_3^{2+}\right)$ and their derivatives.

4.1.1 ECL of Semiconductor QDs

The first QDs ECL behavior was studied by Bard et al. In 2002, they first reported the silicon QDs ECL property [1]. The Si QDs have the ability to store charge in N,N'-dimethylformamide and acetonitrile, which can subsequently lead to light emission upon electron and/or hole transfer. This quality provides electrochemically sensitive optoelectronic properties. In 2006, they observed the ECL emission from silica NPs in aqueous solution [25]. By using $S_2O_8^{2-}$ as the coreactant, octadecyl-protected silica NPs deposited on indium tin oxide (ITO) showed ECL in both anodic and cathodic sweep potentials. In the negative potential scans, the Si NP film could produce a large ECL signal when the potential beyond -0.95 V. The principle was described as follows:

$$S_2O_8^{2-} + e^- \rightarrow SO_4^{2-} + SO_4^{-} \qquad (4.8)$$

$$SO_4^{-} \rightarrow SO_4^{2-} + h^+ \qquad (4.9)$$

$$Si + e^- \rightarrow Si^- \qquad (4.10)$$

$$Si^- + h^+ \rightarrow Si + light \qquad (4.11)$$

The elemental and compound semiconductors, such as Ge [2], CdTe [26], PbS [27], CdSe [28, 29], and ZnS [30], can also generate efficient ECL. The ECL mechanism of semiconductor QDs mainly depends on the annihilation or coreactant ECL reaction. For example, PbS QDs can form oxidized ($R^{\cdot+}$) and reduced ($R^{\cdot-}$) QDs during potential cycling. Two oppositely charged QDs can collide to yield an excited QD in the annihilation process, and the same as CdSe and Ge QDs. Intense ECL signal of CdTe QDs was observed by electron-transfer reaction between reduced NPs and reductively oxidized species of CH_2Cl_2 at negative potential region.

It is believed that the ECL emission is not sensitive to NP size and capping agent used but depends more sensitively on surface chemistry and the presence of surface states [31]. Effective methods are being developed in order to chemically functionalize the surfaces of QDs with suitable ligands or receptors. Ju et al. reported the anodic ECL of MPA-modified water-soluble CdTe QDs in aqueous system. Stable and intensive anodic ECL emission with a peak value at +1.17 V (vs. Ag/AgCl) in pH 9.3 PBS at an ITO electrode was observed. The ECL emission involved the collision between the superoxide ion produced at the ITO surface and the oxidation products of QDs. The whole process of anodic ECL emission was described in the following equations [32]:

$$In/SnO_x + CdTe/CdSR \rightarrow CdTe\left(h^+\right)/CdSR + In/SnO_x\left(e^-\right) \quad (4.12)$$

$$O_2 + In/SnO_x\left(e^-\right) \rightarrow O_2^- + In/SnO_x \quad (4.13)$$

$$CdTe/CdSR + O_2^- \rightarrow CdTe\left(e^-\right)/CdSR + O_2 \quad (4.14)$$

$$CdTe\left(h^+\right)/CdSR + CdTe\left(e^-\right)/CdSR \rightarrow CdTe/CdSR^* \quad (4.15)$$

$$CdTe/CdSR^* \rightarrow CdTe/CdSR + h\nu \quad (4.16)$$

Ju et al. prepared the TGA-protected CdSe QDs for the cathodic ECL research [33]. Chen et al. prepared the CdS-polyamidoamine (CdS-PAMAM) nanocomposite membranes [34] and CdS QDs/CNTs composites [35]. Via in situ electrochemical reduction approach, CdS-PAMAM nanocomposite membranes were prepared on electrode surfaces, and this resulting film showed 55-fold enhanced red ECL compared to that of CdS nanofilm without PAMAM. After combination with CNTs, the CdS QDs/CNTs composites not only enhance their electrochemiluminescent intensity but also decrease their ECL starting potential.

In our group, a series of nanomaterials were synthesized and the corresponding ECL property was investigated. In 2005, we first reported the ECL property of CdS self-assemble nanoparticles. Four ECL peaks were observed in both nonaqueous system and aqueous system [36]. According to the electrochemical behavior of the as-prepared CdS assemblies and the two-equivalent adsorbed surface-state theory, we proposed the mechanism for ECL as annihilation mechanism. Next, we studied the ECL behavior of CdS nanotubes in aqueous solution [37, 38]. By entrapping the CdS nanotubes in carbon paste electrode, two ECL peaks at −0.9 V (vs. Ag/AgCl sat. KCL, *sic passim*, ECL-1), and −1.2 V (ECL-2) were observed when the potential was cycled between 0 and −1.6 V. The ECL-1 was generated via an annihilation process between reduced and oxidized species. The ECL-2 was attributed to the electron-transfer reaction between electrochemically reduced CdS nanocrystal species (CdS$^{\cdot-}$) and coreactants (S$_2$O$_8$$^{2-}$ or H$_2$O$_2$). The corresponding ECL processes were as follows [38]:

S$_2$O$_8$$^{2-}$ as coreactant:

$$CdS\ QDs + ne^- \rightarrow nCdS^{\cdot-} \tag{4.17}$$

$$S_2O_8^{2-} + e^- \rightarrow SO_4^{2-} + SO_4^{\cdot-} \tag{4.18}$$

$$CdS^{\cdot-} + SO_4^{\cdot-} \rightarrow CdS^* + SO_4^{2-} \tag{4.19}$$

$$CdS^* \rightarrow CdS + h\nu \tag{4.20}$$

H$_2$O$_2$ as coreactant:

$$O_2 + H_2O + 2e^- \rightarrow OOH^- + OH^- \tag{4.21}$$

$$2CdS^{\cdot-} + OOH^- + H_2O \rightarrow 3OH^- + 2CdS^* \tag{4.22}$$

$$2CdS^{\cdot-} + H_2O_2 \rightarrow 2OH^- + 2CdS^* \tag{4.23}$$

$$CdS^* \rightarrow CdS + h\nu \tag{4.24}$$

Next, we studied the enhancement ECL effect of 3-aminopropyl-triethoxysilane (APS) on the CdSe QDs in aqueous solution by conjugating APS to the CdSe QDs/carbon nanobute-chitosan (CNT-CHIT) composite film [10]. In the presence of reactive amine groups, APS can facilitate the radical generation and electron-transfer processes during the ECL reaction. After the addition of K$_2$S$_2$O$_8$ as coreactant, APS can catalyze the reaction of CdSe QDs with K$_2$S$_2$O$_8$ based on the following mechanism:

$$CdSe\ QDs + ne^- \rightarrow nCdSe^{\cdot-} \tag{4.25}$$

$$S_2O_8^{2-} + RC_3H_7-NH_2 \rightarrow SO_4^{2-} + SO_4^{\cdot-} + RC_3H_7-NH_2^{+\cdot} \tag{4.26}$$

$$CdSe^{\cdot-} + SO_4^{\cdot-} \rightarrow CdSe^* + SO_4^{2-} \tag{4.27}$$

$$CdSe^* \rightarrow CdSe + h\nu \tag{4.28}$$

However, most of the traditional QDs are made of heavy metal ions (e.g., Cd^{2+}), which may result in potential toxicity that hampers their practical applications. Therefore, systematic cytotoxicity research of QDs is of critical importance for their practical biological and biomedical applications, and a large amount of studies have been carried out for this purpose [39–41]. During the synthesis process, the synthetic methods and surface modifications of QDs will greatly affect their biotoxicity. QDs prepared via the organometallic route and aqueous route were quite different with the surface properties. In contrast to the presence of hydrophobic ligand molecules on the surface of organic QDs, the surface of aqueous QDs are covered with a large amount of hydrophilic molecules [41]. This difference of surface properties could induce distinct cytotoxicity and in vivo behaviors. The surface modifications of QDs could also greatly affect its interaction between the cellular membrane and subsequent uptake into the cells. Taken CdSe as an example, a common surface modification to reduce the cytotoxicity of the core material is coated with a ZnS shell. In one hand, the additional shell semiconductor layer could increase the QDs' photoluminescence. In the other hand, the ZnS shell protects the core CdSe from oxidation and other environmental factors that contribute to cadmium release. Besides, ligands with terminal carboxylic acid, hydroxyl, or amine groups have been used as the charged surface coatings for the QDs protection, which could effectively prevent the core oxidation, cell death, and inflammatory responses.

Another important problem, concerning the body clearance of these nanoparticles, is attracted more and more attention. When employing living mouse for the in vivo imaging study by QDs injection, Ballou et al. [42] found that methoxy-terminated poly(ethylene glycol) amine QDs (mPEG-QDs) remained for at least one month in liver, lymph nodes, and bone marrow. Therefore, the use of QDs in vivo must be critically examined.

Considering these biotoxicities, various new kinds of QDs are emerged in recent years such as silicon QD [43], carbon dot [44], graphene QD [45]. Owing to their special cadmium free property, excellent biocompatibility, and environmentally friendliness, these novel nanomaterials gained significant consideration after being successfully prepared.

4.1.2 ECL of GQDs

As new type of QD, graphene QDs has been widely studied nowadays. In 2008, Bard's group [46] first reported the ECL from electrochemically oxidized highly oriented pyrolytic graphite (HOPG) and from a suspension of graphene oxide platelets, whose results were presented in Fig. 4.2. They supposed that the smaller aromatic hydrocarbon-like domains formed on the "graphitic" layers by interruption of the conjugation could form the oxidized and emitting centers. In the case of individual graphene oxide NPs, ECL signal could also be detected by using tri-n-propylamine (TPrA) as a coreactant at relatively high concentration. At positive potential of ~1.15 V, oxidation of graphene oxide NPs takes place either directly on the working electrode during collision or via TPrA radical cations. The highly reductive radical intermediate could be generated by the deprotonation reaction of TPrA, and the excited state of graphene oxide for the generation of ECL could be formed between the radiative recombination of this radical and the graphene oxide NP.

Afterward, luminescent property of graphene was studied widely. In 2011, Li and co-workers [47] observed the cathodic ECL response of luminol at a positive potential with a strong light emission on a graphene-modified glassy carbon electrode. With the utilization of the excellent electrocatalytic properties, graphene could facilitate the reduction in O_2 in a solution dissolved with trace oxygen, which was critical on the cathodic ECL behavior of luminol on the GR-CHIT/GC

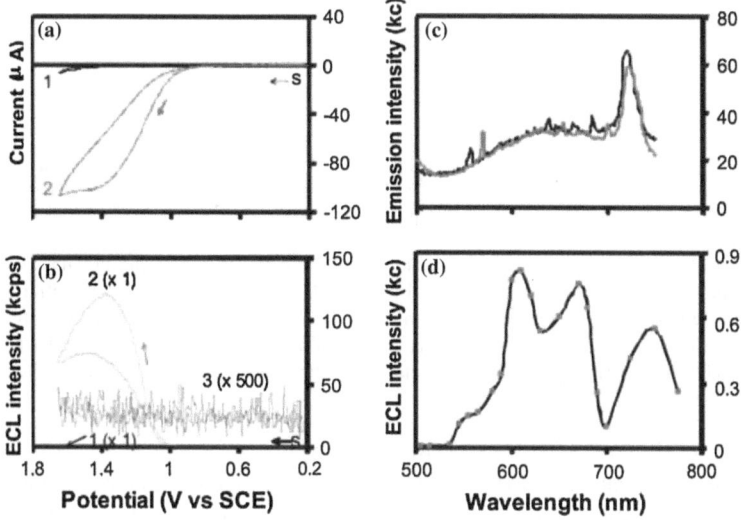

Fig. 4.2 CV (**a**) and ECL intensity versus potential (**b**): (*black*) 0.1 M NaClO$_4$ and PBS only; (*red*) with addition of 13 mM TPrA; (*blue*) expanded scale of ECL intensity of curve 1. Potential scan rate = 20 mV/s at an HOPG electrode (area = 0.07 cm^2). **c** Photoluminescence in air of an HOPG electrode after ECL experiments. Excitation wavelength is 430 nm (*black*) and 440 nm (*red*). **d** ECL spectrum of HOPG in 0.1 M NaClO$_4$, PBS (pH 7.0) containing 13 mM TPrA. Reprinted with permission from Ref. [46]. Copyright 2009 American Chemical Society

electrodes. Similar phenomenon was observed by Yuan's group [48] using hemin–graphene nanosheets (H-GNs) as the ECL amplification. Owing to the superior electrical conductivity of H-GNs, they were able to promote electron transfer, so as to amplify the luminol ECL signals of the prepared biosensor.

In the author's group, a series of graphene QD (GQD) and its nanocomposites were synthesized and studied. Via the electrostatic interactions between negatively charged thioglycolic acid (TGA)-modified CdSe QDs and positively poly (diallyldimethylammonium chloride) (PDDA)-protected graphene, the P-GR-CdSe composites were successfully prepared and used to construct an ECL immunosensor [49]. With the help of the excellent conductivity, extraordinary electron-transport properties and large specific surface area, the interfusion of PDDA-protected graphene (P-GR) with CdSe QDs film not only improved the ECL intensity, response speed, and stability, but also held high levels of protein loading, which resulted in extreme sensitivity.

Later, we prepared the greenish-yellow luminescent graphene quantum dots (gGQDs) with a quantum yield (QY) up to 11.7 % through the assistance of microwave irradiation [50]. As shown in Fig. 4.3, ECL is observed from the gGQDs for the first time in 0.05 M, pH 7.4 Tris–HCl buffer solution (TBS) with 0.1 M K$_2$S$_2$O$_8$ as coreactant. The possible ECL mechanism was proposed as follows: firstly, strongly oxidizing SO$_4^-$ radicals and GQDs$^-$ radicals were produced by electrochemical reduction in S$_2$O$_8^{2-}$ and GQDs, respectively.

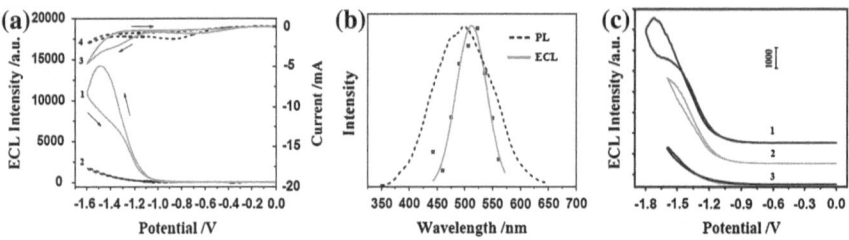

Fig. 4.3 a ECL–potential curves and cyclic voltammograms (*CVs*) of the gGQDs (*1, 3*) and background (*2, 4*) with concentration of 20 ppm in 0.05 M Tris–HCl (pH 7.4) buffer solution containing 0.1 M $K_2S_2O_8$. Scan rate: 100 mV s^{-1}. **b** PL (λex = 340 nm) and ECL spectra for the gGQDs-$K_2S_2O_8$ system. **c** ECL–potential *curves of the background* (*1*) and bGQDs (20 ppm) (*2, 3*) in 0.05 M Tris–HCl (pH 7.4) buffer solution containing 0.1 M $K_2S_2O_8$. Reprinted with permission from Ref. [50]. Copyright 2012 Wiley

Then, SO_4^- radicals could react with $GQDs^{\cdot-}$ via electron-transfer annihilation, producing an excited state (GQDs*) that finally emitted light, which could be described as the following equations:

$$GQDs + ne^- \rightarrow nGQDs^{\cdot-} \tag{4.29}$$

$$S_2O_8^{2-} + e^- \rightarrow S_2O_8^{\cdot2-} \tag{4.30}$$

$$S_2O_8^{\cdot2-} \rightarrow SO_4^{2-} + SO_4^{\cdot-} \tag{4.31}$$

$$GQDs^{\cdot-} + SO_4^{\cdot-} \rightarrow GQDs^* + SO_4^{2-} \tag{4.32}$$

$$GQDs^* \rightarrow GQDs + h\nu \tag{4.33}$$

4.2 QDs ECL for DNA Biosensing

The combination for the sensitive ECL detection with extremely selective biological interaction between DNA/aptamer-probe assays has attracted more and more interest over the past years. ECL allows the detection of analytes at low concentrations over a wide linear range. What is more, ECL could combine with many other analytical methods such as high-performance liquid chromatography (HPLC), liquid chromatography (LC), capillary electrophoresis (CE), and flow injection analysis (FIA).

4.2.1 QDs ECL for DNA Analysis

QD ECL technique was widely used in the field of DNA biosensing. Various QDs of CdTe [51, 52], CdS [53, 54], and their composites [55–59] were employed as the ECL luminants. In 2009, with CdS:Mn NCs as the ECL luminophores and Au NPs functioning as both ECL quencher and enhancer, Xu's group [59] demonstrated a simple ECL sensing platform for highly sensitive and specific detection of DNA targets, which was illustrated in Fig. 4.4. Due to the Förster resonance energy transfer (FRET) between the CdS:Mn NC film and AuNPs, the quenching of QD ECL could be produced when the QDs and AuNPs were at close proximity. And ECL enhancement was obtained after the interaction of the excited CdS:Mn NCs with ECL-induced surface plasmon resonance in AuNPs at large separation. Considering that graphene owns the excellent electrochemical property, they lately synthesized the K-doped graphene–CdS:Eu NC composites as the ECL emitter [58]. Via the electrostatic interactions between negatively charged 3-mercaptopropionic acid (MPA)-modified CdS:Eu NCs and positively charged graphene, the novel K-doped graphene–CdS:Eu NC composites were prepared. By utilization of the facts that K-doped graphene could enhance the rate of electron transfer and that Eu^{3+} ions could alter the surface of CdS NCs to create a new surface-state-$Eu3^+$ complex, the NC-ECL performance exhibited a large enhancement in the ECL intensity. The layer-by-layer assembly K-GR–NC composites film on the glassy carbon electrode (GCE) not only improved the ECL intensity, but also provided a large specific surface for high levels of DNA loading, which resulted in extreme sensitivity with a detection limit of 50 aM.

For the purpose of improving the sensitivity, Jin et al. [51] employed the nanoporous gold leaf (NPGL) electrode as the working electrode for DNA detection using CdTe QDs as the labels. Due to these ultrathin nanopores, sensitivity of the ECL biosensor is remarkably increased. In this assay, target DNA (t-DNA) was hybridized with capture DNA (c-DNA) bound on the NPGL electrode, and amino-modified probe DNA was hybridized with the t-DNA to yield the sandwich hybrids on the NPGL electrode. At last, MPA-capped CdTe QDs are labeled to the amino group end of the sandwich hybrids as the signal producer in the presence of $S_2O_8^{2-}$ as coreactant. The maximum ECL intensity on the curve is proportional to

Fig. 4.4 DNA ECL sensing platform based on energy transfer between CdS:Mn NCs and AuNPs. Reprinted with permission from Ref. [59]. Copyright 2009 Royal Society of Chemistry

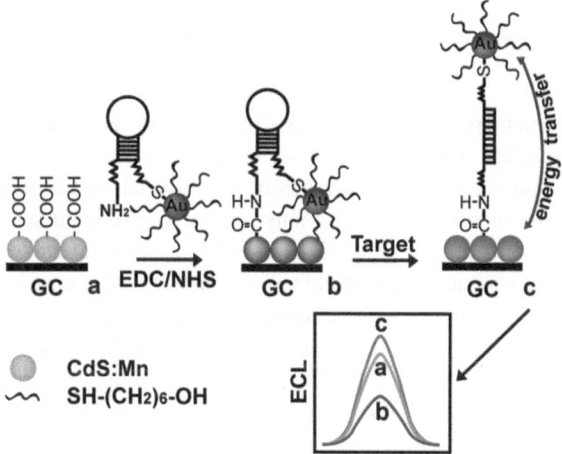

Table 4.1 QDs ECL employed in the field of DNA biosensing

QD	Coreactant	Linear range (mol L^{-1})	References
CdTe	$S_2O_8^{2-}$	5×10^{-11}–1×10^{-15}	Hu et al. [51]
CdTe	O_2	1×10^{-10}–5×10^{-15}	Deng et al. [52]
CdS	$S_2O_8^{2-}$	1×10^{-16}–5×10^{-18}	Zhou et al. [53]
CdS	$S_2O_8^{2-}$	1×10^{-10}–5×10^{-18}	Divsara and Ju [54]
CdSe@ZnS	$S_2O_8^{2-}$	5×10^{-6}–5×10^{-9}	Huang et al. [55]
CdTe@ZnO	$S_2O_8^{2-}$	10^{-14}–10^{-19}	Liu et al. [56]
CdTe@SiO_2	$S_2O_8^{2-}$	2×10^{-6}–1×10^{-10}	Wei et al. [57]
CdS:Eu	H_2O_2	1×10^{-11}–5×10^{-17}	Zhou et al. [58]
CdS:Mn	$S_2O_8^{2-}$	5×10^{-17}–5×10^{-15}	Shan et al. [59]

t-DNA concentration with a linear range of 5×10^{-15}–1×10^{-11} mol L^{-1}. Given that the Cd-component QDs cytotoxicity could be caused by the release of heavy metal ions and the instability in harsh environments, Liu et al. [57] synthesized the CdTe@SiO2 composite through a reverse microemulsion method. The prepared products not only retained high fluorescence intensity but also avoided cytotoxicity due to the protection by the SiO_2 shell. DNA detection was realized by outputting a remarkable ECL signal of the CdTe@SiO_2 labeled to the probe DNA with a low detection limit of 0.03 nM and a wide dynamic range from 0.1 nM to 2 mM.

In the author's group, we [55] developed a simple and convenient assay with QDs as the labels for DNA detection. Different from the common sandwich-type strategy, in this system, the target DNA oligonucleotides are directly hybridized with the probe DNA oligonucleotides for DNA detection, which reduces the hybridization process, thus it is relatively simple and time-saving. After the QDs bind to the target DNA via the biotin–avidin system, the DNA hybridization event can be detected by ECL signal.

Table 4.1 shows the QDs ECL in DNA biosensing.

4.2.2 QDs ECL for Aptasensor Analysis

Aptamers are a new class of single-stranded DNA/RNA molecules, which are selected from synthetic nucleic acid libraries via the selection procedure called systematic evolution of ligands by exponential enrichment (SELEX). Since first discovered in 1990s [60, 61], many aptamers have been selected for corresponding targets combination ranging from metal ions, organic molecules, biomolecules, to entire organism and even whole cells. Aptamers offer remarkable convenience in design and modification of their structures, which has motivated them to generate a great variety of aptamer sensors (aptasensors) that exhibit high sensitivity as well as specificity.

Based on the QDs ECL, a series of aptamer biosensors have been constructed and applied for the detection of various targets ranging from small biomolecule to entire cell, which were listed in Table 4.2.

Table 4.2 QDs ECL employed in the field of Aptamer biosensing

QD	Target	Linear range	References
CdSe/ZnS	ATP	0.018–90.72 mM	Huang et al. [62]
CdSe–CdS	ATP	1.0×10^{-8}–8.0×10^{-7} M	Jie et al. [63]
TiO$_2$–N	Adenosine	10 nM–1.0 mM	Tian et al. [64]
CdSe–ZnS	Cancer cell	4×10^2–10^4 cells mL^{-1}	Jie et al. [65]
CdS-ZnS	Cancer cell	60–1,000 cells	Jie et al. [66]
CdSe/ZnS	Lysozyme	/	Huang et al. [67]
CdSe/ZnS	Thrombin	27.2–545 nM	Chen et al. [68]
CdSe/ZnS	Thrombin	0.01–50 nM	Xie et al. [69]
CdSe	Thrombin	0–64 μg mL^{-1}	Li et al. [70]
CdTe/SiO2	Thrombin	5.0 aM–5.0 fM	Shan et al. [71]
CdTe	Thrombin	0.5–800 pM	Huang and Zhu [72]
CdS	Thrombin	1.0×10^{-15}–1.0×10^{-13} M	Guo et al. [73]
CdTe	Pb^{2+}	2.0×10^{-10}–1.0×10^{-8} M	Hai et al. [74]

In our group, using the QDs as the signal tranducers, a series of aptasensors were developed for the detection of lysozyme [67], thrombin [72], and adenosine 5′-triphosphate (ATP) [62]. Through the aptamer-target-specific affinity and the rules of Watson–Crick base pairing, aptasensors for lysozyme and ATP detection were fabricated, respectively.

Take ATP detection as an example [62], after the thiol modified anti-ATP probes were immobilized onto the pretreated Au electrode, ATP solution was dropped for the formation of aptamer–ATP bioaffinity complexes. The biotin-modified complementary DNA (biotin-cDNA, in terms of the probe) oligonucleotides were hybridized with the remnant-free probes. At last, the avidin-modified QDs were bound to the aptasensor through the biotin–avidin system in the existence of biotin-cDNA. The ECL intensity, as the readout signal for the aptasensor, was responsive to the amount of QDs bonded to the cDNA oligonucleotides, which was inversely proportional to the combined target analyte indirectly. Later, Jie et al. [63] employed superstructural dendrimeric CdSe–CdS–QDs as the emitter and successfully applied to amplified ECL assays of ATP using DNA cycle amplification technique. The whole fabrication process was given in Fig. 4.5. Compared with the pure QDs, the superstructure exhibits highly enhanced ECL signal and could be easily labeled, separated, and immobilized onto a magnetic electrode.

In the case that thrombin owned two different aptamers, we ingeniously grafted the concept of sandwich immunoreaction into the aptamer field and constructed a sandwich aptasensor by using QDs ECL technique for thrombin detection [72]. The thiol-terminated aptamer with 15 nucleotides (probe I) was first immobilized on Au electrode, and then, thrombin and another 5′-biotin-modified aptamer (29 nucleotides, probe II) were incubated to the above electrode, respectively, in order to form the sandwich structure of probe I/thrombin/probe II. Streptavidin-modified QDs (avidin–QDs) were bound to probe II via the biotin–avidin system. Thrombin was

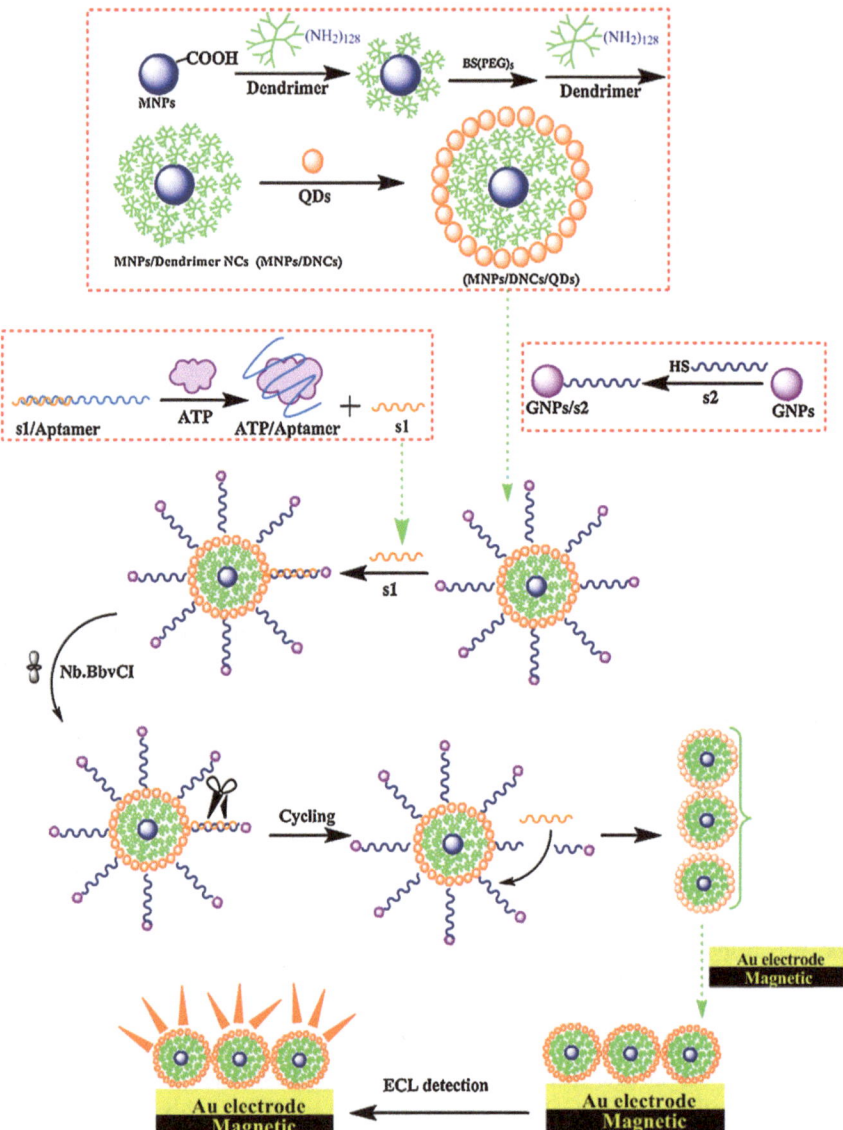

Fig. 4.5 Schematic diagram for the biosensor fabrication for ATP detection. Reprinted with permission from Ref. [63]. Copyright 2011 Elsevier B.V

detected by measurement of the ECL intensity of the bound QDs. It provided an alternative convenient, low-cost, and specific method for protein detection. What is more, the application of QDs ECL in aptasensors may intrigue new interest in the investigation of the QDs ECL and promote the exploitation in the bioapplications. Based on this strategy, Yuan's group [68] prepared new ECL signal amplification labels through layer-by-layer (LBL) assembly of CdTe QDs onto the surfaces of the polystyrene (PS) microbeads. Instead of the conventional single CdTe QD one, numerous CdTe QDs were involved in each molecular recognition event, which were illuminated in Fig. 4.6. The analytical signal output was thus expected to be dramatically amplified. With this new PS-(CdTe)$_2$ labels, about 2–4 orders of magnitude improvement in detection limit for thrombin is obtained compared with that of other universal signal amplification routes. The detection limit could be pushed down further by increasing the number of CdTe QD layer coated on the PS beads.

Besides the traditional QDs, Xu's group [64] reported the highly enhanced ECL from a nitrogen doped TiO$_2$ nanotube array. They found that the nitrogen incorporated in TiO$_2$ NTs could greatly enhance the ECL intensity and make the ECL spectrum red shift. Using K$_2$S$_2$O$_8$ as the coreactant, the TiO$_2$–N NTs electrode can enhance the ECL intensity by 10.6-fold and move the onset ECL potential more positively by about 400 mV as compared to the pure TiO$_2$ NTs electrode. And the possible mechanism was forwarded as following:

$$S_2O_8{}^{2-} + e \rightarrow SO_4{}^{2-} + SO_4{}^{\cdot-} \tag{4.34}$$

$$SO_4{}^{\cdot-} \rightarrow SO_4{}^{2-} + h^+ \tag{4.35}$$

$$TiO_2 - N + h^+ \rightarrow N - TiO_2(h^+) \tag{4.36}$$

$$N-TiO_2(h^+) + e^- \rightarrow (N-TiO_2)^* \tag{4.37}$$

$$(N-TiO_2)^* \rightarrow N-TiO_2 + h\lambda \tag{4.38}$$

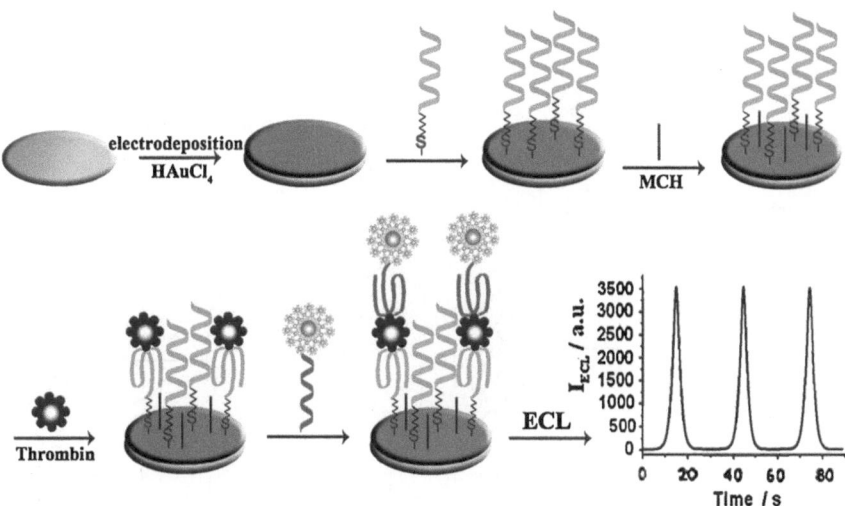

Fig. 4.6 Schematic presentation of the PS-(CdTe)$_2$ assembly-labeled ECL detection of thrombin. Reprinted with permission from Ref. [68]. Copyright 2011 Elsevier

When applied this nanomaterial to develop an ultrasensitive ECL aptasensor for the detection of adenosine in cancer cells, the ECL signal linearly increases with the increase in the logarithm of the ATP concentration over the range 10 nM–1.0 mM with a limit detection of around 10 nM.

Not only could the small biomolecules, even the entire cells be detected via the aptamer-based ECL array. In 2011, Jie and coworkers [65] prepared a novel dendrimer/CdSe/ZnS quantum dot nanocluster (NC) and used as an ECL probe for versatile assays of cancer cells. As given in Fig. 4.7, a large number of CdSe-ZnS-QDs could be labeled on the NCs due to the many functional amine groups within the NCs, which could significantly amplify the QD's ECL signal. The ECL biosensor for cancer cells was directly accomplished by using the biobarcode technique to avoid cross-reaction. Moreover, the utilization of magnetic beads (MBs) for aptamers immobilization greatly simplified the separation procedures and favored for the sensitivity improvement. The high specificity of aptamers to target cells and the biobarcode technique to avoid cross-reaction endow the biosensor with high selectivity and sensitivity. The changes of ECL peak intensity increased linearly with the cell concentrations in the range of 400–10,000 cells mL^{-1}, and

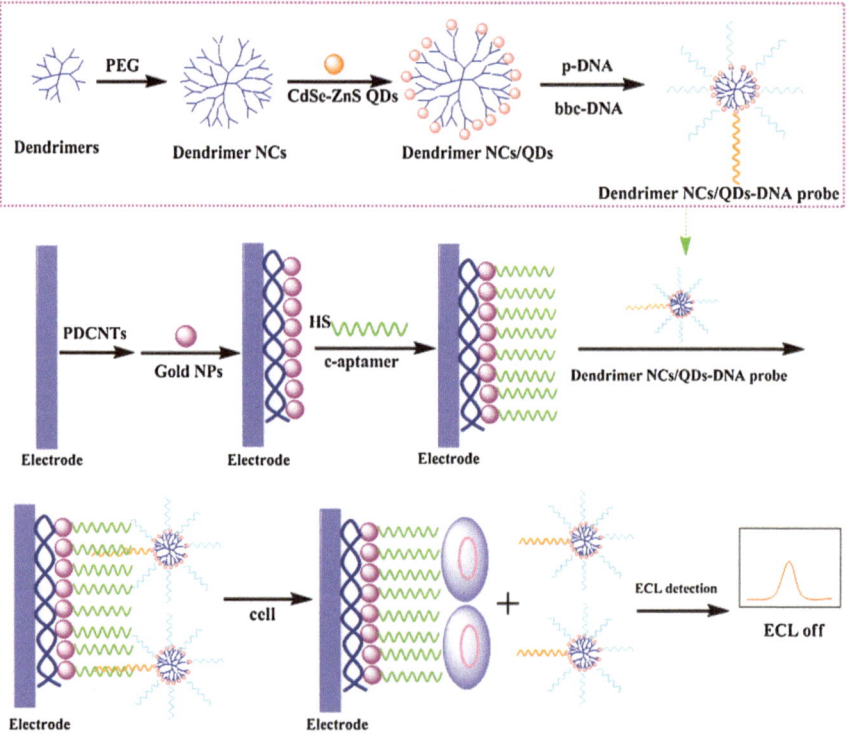

Fig. 4.7 Fabrication steps of the dendrimer NCs/QDs-DNA probe and ECL biosensor for signal-off detection of cells. Reprinted with permission from Ref. [65]. Copyright 2011 American Chemical Society

the detection limit was calculated to be 210 cells mL^{-1} at 3σ. Afterward, they [66] employed another type of dendritic CdSe/ZnS-QDs nanocomposite to construct the ECL aptasensor for the cancer cells detection based on ECL quenching of QDs by gold nanoparticles (NPs). DNA conjugation and gold NPs linking and sensing target cells can be directly performed on the magnetic nanocomposites for cell detection based on ECL quenching.

References

1. Ding Z, Quinn BM, Haram SK, Pell LE, Korgel BA, Bard AJ (2002) Electrochemistry and electrogenerated chemiluminescence from silicon nanocrystal quantum dots. Science 296:1293–1297
2. Myung N, Lu X, Johnston KP, Bard AJ (2004) Electrogenerated Chemiluminescence of Ge Nanocrystals. Nano Lett 4:183–185
3. Hercules DM (1964) Chemiluminescence resulting from electrochemically generated species. Science 145:808–809
4. ViscoR E, Chandross EA (1964) Electroluminescence in solutions of aromatic hydrocarbons. J Am Chem Soc 86:5350–5351
5. Santhanam KS, Bard AJ (1965) Chemiluminescence of electrogenerated 9,10-diphenylanthracene anion radical. J Am Chem Soc 87:139–140
6. Bertolino C, MacSweeney M, Tobin J, O'Neill B, Sheehan MM, Coluccia S, Berney H (2005) A monolithic silicon based integrated signal generation and detection system for monitoring DNA hybridisation. Biosens Bioeletron 21:565–573
7. Zhan W, Bard AJ (2007) Electrogenerated chemiluminescence 83 immunoassay of human c-reactive protein by using Ru(bpy)$_3$$^{2+}$-encapsulated liposomes as labels. Anal Chem 79:459–463
8. Ryoji K, Kumi A, Kohei N, Dai K, Osamu N (2010) Development of electrogenerated chemiluminescence-based enzyme linked immunosorbent assay for Sub-pM detection. Anal Chem 82:1692–1697
9. Liu X, Ju HX (2008) Coreactant enhanced anodic electrochemiluminescence of CdTe quantum dots at low potential for sensitive biosensing amplified by enzymatic cycle. Anal Chem 80:5377–5382
10. Jie GF, Zhang JJ, Wang DC, Cheng C, Chen HY, Zhu JJ (2008) Electrochemiluminescence immunosensor based on CdSe nanocomposites. Anal Chem 80:4033–4039
11. Rivera VR, Gamez FJ, Keener WK, Poli MA (2006) Rapid detection of clostridium botulinum toxins A, B, E, and F in clinical samples, selected food matrices, and buffer using paramagnetic bead-based electrochemiluminescence detection. Anal Biochem 353:248–256
12. Luo LR, Zhang ZJ, Chen LJ, Ma LF (2006) Chemiluminescent imaging detection of staphylococcal enterotoxin C1 in milk and water samples. Food Chem 97:355–360
13. Bruno JG, Kiel JL (1999) In vitro selection of DNA aptamers to anthrax spores with electrochemiluminescence detection. Biosens Bioelectron 14:457–464
14. Wolter A, Niessner R, Seidel M (2008) Detection of *Escherichia coli* O157:H7, *Salmonella typhimurium*, and *Legionella pneumophila* in Water Using a flow-through chemiluminescence microarray readout system. Anal Chem 80:5854–5863
15. Zhang J, Qi HL, Li Y, Yang J, Gao Q, Zhang CX (2008) Electrogenerated chemiluminescence DNA biosensor based on hairpin DNA probe labeled with ruthenium complex. Anal Chem 80:2888–2894
16. Huang HP, Li JJ, Zhu JJ (2011) Electrochemiluminescence based on quantum dots and their analytical application. Anal Methods 3:33–42

17. Hu LZ, Bian Z, Li HJ, Han S, Yuan YL, Gao LX, Xu GB (2009) [Ru(bpy)$_2$dppz]$^{2+}$ electro-chemiluminescence switch and its applications for DNA interaction study and label-free ATP aptasensor. Anal Chem 81:9807–9811

18. Yin XB, Xin YY, Zhao Y (2009) Label-Free electrochemiluminescent aptasensor with atto-molar mass detection limits based on a Ru(phen))$_3$$^{2+}$-double-strand DNA composite film electrode. Anal Chem 81:9299–9305

19. Kulmala S, Ala-Kleme T, Vare L, Helin M, Lehtinen T (1999) Hot electron-induced elec-trogenerated luminescence of Tl(I) at disposable oxide-covered aluminum electrodes. Anal Chim Acta 398:41–47

20. Kankare J, Falden K, Kulmala S, Haapakka K (1992) Cathodically induced time-resolved lanthanide(III) electroluminescence at stationary aluminium disc electrodes. Anal Chim Acta 256:17–28

21. Kankare J, Haapakka K, Kulmala S, Nanto V, Eskola J, Takalo H (1992) Immunoassay by time-resolved electrogenerated luminescence. Anal Chim Acta 266:205–212

22. Miao W (2007) Electrogenerated chemiluminescence. In: Zoski CG (ed). Handbook of elec-trochemistry. Elsevier, The Netherlands

23. MChang M, Saji T, Bard AJ (1977) Electrogenerated chemiluminescence 30 electrochemical oxidation of oxalate ion in the presence of luminescers in acetonitrile solutions. J Am Chem Soc 99:5399–5403

24. Rubinstein I, Bard AJ (1981) Electrogenerated chemiluminescence 37 Aqueous ecl systems based on tris(2,2'-bipyridine) ruthenium(2+) and oxalate or organic acids. J Am Chem Soc 103:512–516

25. Bae Y, Lee DC, Rhogojina EV, Jurbergs DC, Korgel BA, Bard AJ (2006) Electrochemistry and electrogenerated chemiluminescence of films of silicon nanoparticles in aqueous solu-tion. Nanotechnology 17:3791

26. Bae Y, Myung N, Bard AJ (2004) Electrochemistry and electrogenerated chemiluminescence of CdTe nanoparticles. Nano Lett 4:1153–1161

27. Sun LF, Bao L, Hyun BR, Bartnik AC, Zhong YW, Reed JC, Pang DW, Abruña HD, Malliaras GG, Wise FW (2009) Electrogenerated chemiluminescence from PbS quantum dots. Nano Lett 9:789–793

28. Myung N, Ding Z, Bard AJ (2002) Electrogenerated chemiluminescence of CdSe nanocrys-tals. Nano Lett 2:1315–1319

29. Myung N, Bae Y, Bard AJ (2003) Effect of surface passivation on the electrogenerated chemiluminescence of CdSe/ZnSe nanocrystals. Nano Lett 3:1053–1055

30. Shen L, Cui X, Qi H, Zhang C (2007) Electrogenerated chemiluminescence of ZnS nanopar-ticles in alkaline aqueous solution. J Phys Chem C 111:8172–8175

31. Bard AJ, Ding Z, Myung N (2005) Electrochemistry and electrogenerated chemilumines-cence of semiconductor nanocrystals in solutions and in films. Struct Bond 118:1–57

32. Liu X, Jiang H, Lei JP, Ju HX (2007) Anodic electrochemiluminescence of CdTe quantum dots and its energy transfer for detection of catechol derivatives. Anal Chem 79:8055–8060

33. Jiang H, Ju HX (2007) Enzyme–quantum dots architecture for highly sensitive electrochemi-luminescence biosensing of oxidase substrates. Chem Commun 4:404–406

34. Lu C, Wang XF, Xu JJ, Chen HY (2008) Electrochemical modulation of electrogenerated chemiluminescence of CdS nano-composite. Electrochem Commun 10:1530–1532

35. Ding SN, Xu JJ, Chen HY (2006) Enhanced solid-state electrochemiluminescence of CdS nanocrystals composited with carbon nanotubes in H$_2$O$_2$ solution. Chem Commun 34:3631–3633

36. Ren T, Xu JZ, Tu YF, Xu S, Zhu JJ (2005) Electrogenerated chemiluminescence of CdS spherical assemblies. Electrochem Commun 7:5–9

37. Miao JJ, Ren T, Lin D, Zhu JJ, Chen HY (2005) Double-template synthesis of CdS nano-tubes with strong electrogenerated chemiluminescence. Small 1:802–805

38. Jie GF, Liu B, Miao JJ, Zhu JJ (2007) Electrogenerated chemiluminescence from CdS nano-tubes and its sensing application in aqueous solution. Talanta 71:1476–1480

39. Kirchner C, Liedl T, Kudera S, Pellegrino T, Javier AM, Gaub HE, Stölzle S, Fertig N, Parak WJ (2005) Cytotoxicity of colloidal CdSe and CdSe/ZnS nanoparticles. Nano Lett 5:331–338
40. Green M, Howman E (2005) Semiconductor quantum dots and free radical induced DNA nicking. Chem Commun 1:121–123
41. Chen N, He Y, Su YY, Li XM, Huang Q, Wang HF, Zhang XZ, Tai RZ, Fan CH (2012) The cytotoxicity of cadmium-based quantum dots. Biomater 33:1238–1244
42. Ballou B, Lagerholm BC, Ernst LA, Bruchez MP, Waggoner AS (2004) Noninvasive imaging of quantum dots in mice. Bioconjugate Chem 15:79–86
43. Li ZF, Ruckensteln E (2004) Water-soluble poly (acrylic acid) grafted luminescent silicon nanoparticles and their use as fluorescent biological staining labels. Nano Lett 4:1463–1467
44. Sun YP, Zhou B, Lin Y, Wang W, Fernando KAS, Pathak P, Meziani MJ, Harruff BA, Wang X, Wang HF, Luo PG, Yang H, Kose ME, Chen B, Veca LM, Xie SY (2006) Quantum-sized carbon dots for bright and colorful photoluminescence. J Am Chem Soc 128:7756–7757
45. Shen JH, Zhu YH, Yang XL, Li CZ (2012) Graphene quantum dots: emergent nanolights for bioimaging, sensors, catalysis and photovoltaic devices. Chem Commun 48:3686–3699
46. Fan FR, Park SJ, Zhu YW, Ruoff RS, Bard AJ (2009) Electrogenerated chemiluminescence of partially oxidized highly oriented pyrolytic graphite surfaces and of graphene oxide nano-particles. J Am Chem Soc 131:937–939
47. Xu SJ, Liu Y, Wang TH, Li JH (2011) Positive potential operation of a cathodic electrogen-erated chemiluminescence immunosensor based on luminol and graphene for cancer bio-marker detection. Anal Chem 83:3817–3823
48. Zhang MH, Yuan R, Chai YQ, Chen SH, Zhong X, Zhong HA, Wang C (2012) A cathodic electrogenerated chemiluminescence biosensor based on luminol and hemin-graphene nanosheets for cholesterol detection. RSC Adv 2:4639–4641
49. Li LL, Liu KP, Yang GH, Wang CH, Zhang JR, Zhu JJ (2011) Fabrication of graphene–quan-tum dots composites for sensitive electrogenerated chemiluminescence immunosensing. Adv Funct Mater 21:869–878
50. Li LL, Ji J, Fei R, Wang CZ, Lu Q, Zhang JR, Jiang LP, Zhu JJ (2012) A facile microwave avenue to electrochemiluminescent two-color graphene quantum dots. Adv Funct Mater 22:2971–2979
51. Hu XF, Wang RY, Ding Y, Zhang XL, Jin WR (2010) Electrochemiluminescence of CdTe quantum dots as labels at nanoporous gold leaf electrodes for ultrasensitive DNA analysis. Talanta 80:1737–1743
52. Deng SY, Cheng LX, Lei JP, Cheng Y, Huang Y, Ju HX (2013) Label-free electrochemilumi-nescent detection of DNA by hybridization with a molecular beacon to form hemin/G-quad-ruplex architecture for signal inhibition. Nanoscale 5:5435–5441
53. Zhou H, Liu J, Xu JJ, Chen HY (2011) Ultrasensitive DNA detection based on Au nanopar-ticles and isothermal circular double-assisted electrochemiluminescence signal amplification. Chem Commun 47:8358–8360
54. Divsara F, Ju HX (2011) Electrochemiluminescence detection of near single DNA molecules by using quantum dots–dendrimer nanocomposites for signal amplification. Chem Commun 47:9879–9881
55. Huang HP, Li JJ, Tan YL, Zhou JJ, Zhu JJ (2010) Quantum dot-based DNA hybridization by electrochemiluminescence and anodic stripping voltammetry. Analyst 135:1773–1778
56. Liu F, Liu H, Zhang M, Yu JH, Wang SW, Lu JJ (2013) Ultrasensitive electrochemilumi-nescence detection of lengthy DNA molecules based on dual signal amplification. Analyst 138:3463–3469
57. Wei W, Zhou J, Li HN, Yin LH, Pu YP, Liu SQ (2013) Fabrication of CdTe@SiO$_2$ nano-probes for sensitive electrogenerated chemiluminescence detection of DNA damage. Analyst 138:3253–3258
58. Zhou H, Zhang YY, Liu J, Xu JJ, Chen HY (2013) Efficient quenching of electrochemilu-minescence from K-doped graphene–CdS: Eu NCs by G-quadruplex–hemin and target recy-cling-assisted amplification for ultrasensitive DNA biosensing. Chem Commun 49:2246–2248

59. Shan Y, Xu JJ, Chen HY (2009) Distance-dependent quenching and enhancing of electro-chemiluminescence from a CdS: Mn nanocrystal film by Au nanoparticles for highly sensitive detection of DNA. Chem Commun 45:905–907
60. Ellington AD, Szostak JW (1990) In vitro selection of RNA molecules that bind specific ligands. Nature 346:818–822
61. Tuerk C, Gold L (1990) Systematic evolution of ligands by exponential enrichment: RNA ligands to bacteriophage T4 DNA polymerase. Science 249:505–510
62. Huang HP, Tan YL, Shi JJ, Liang GX, Zhu JJ (2010) DNA aptasensor for the detection of ATP based on quantum dots electrochemiluminescence. Nanoscale 2:606–612
63. Jie GF, Yuan JX, Zhang J (2012) Quantum dots-based multifunctional dendritic superstructure for amplified electrochemiluminescence detection of ATP. Biosens Bioelectron 31:69–76
64. Tian CY, Xu JJ, Chen HY (2012) A novel aptasensor for the detection of adenosine in cancer cells by electrochemiluminescence of nitrogen doped TiO$_2$ nanotubes. Chem Commun 48:8234–8236
65. Jie GF, Wang L, Yuan JX, Zhang SS (2011) Versatile electrochemiluminescence assays for cancer cells based on dendrimer/CdSe–ZnS–quantum dot nanoclusters. Anal Chem 83:3873–3880
66. Jie GF, Yuan JX, Huang TY, Zhao YB (2012) Electrochemiluminescence of dendritic magnetic quantum dots nanostructure and its quenching by gold nanoparticles for cancer cells assay. Electroanal 24:1220–1225
67. Huang HP, Jie GF, Cui RJ, Zhu JJ (2009) DNA aptamer-based detection of lysozyme by an electrochemiluminescence assay coupled to quantum dots. Electrochem Commun 11:816–818
68. Chen Y, Jiang BY, Xiang Y, Chai YQ, Yuan R (2011) Aptamer-based highly sensitive electro-chemiluminescent detection of thrombin via nanoparticle layer-by-layer assembled amplification labels. Chem Commun 47:7758–7760
69. Xie LL, You LQ, Cao XY (2013) Signal amplification aptamer biosensor for thrombin based on a glassy carbon electrode modified with graphene, quantum dots and gold nanoparticles. Spectrochim Acta Part A: Mol Biomol Spec 109:110–115
70. Li YF, Liu LL, Fang XL, Bao JC, Han M, Dai ZH (2012) Electrochemiluminescence biosensor based on CdSe quantum dots for the detection of thrombin. Electrochim Acta 65:1–6
71. Shan Y, Xu JJ, Chen HY (2011) Enhanced electrochemiluminescence quenching of CdS: Mn nanocrystals by CdTe QDs-doped silica nanoparticles for ultrasensitive detection of thrombin. Nanoscale 3:2916–2923
72. Huang HP, Zhu JJ (2009) DNA aptamer-based QDs electrochemiluminescence biosensor for the detection of thrombin. Biosens Bioelectron 25:927–930
73. Guo YS, Jia XP, Zhang SS (2011) DNA cycle amplification device on magnetic microbeads for determination of thrombin based on graphene oxide enhancing signal-on electrochemilu-minescence. Chem Commun 47:725–727
74. Hai H, Yang F, Li JP (2013) Electrochemiluminescence sensor using quantum dots based on a G-quadruplex aptamer for the detection of Pb^{2+}. RSC Adv 3:13144–13148

Chapter 5
Quantum Dot-Electrochemical and Photoelectrochemical Biosensing

Abstract Besides excellent optical and electrochemiluminescence properties of quantum dots, their electrochemical and photoelectrochemical behaviors have also been discovered and well employed in DNA biosensing. This chapter presents a general description of the electrochemical and photoelectrochemical properties of QDs with their applications in DNA biosensing.

Keywords Electrochemical labels • Photoelectrochemical analysis • Quantum dots • DNA biosensing • Aptamer analysis

5.1 QDs as Electrochemical Labels

5.1.1 The Electrochemical Behavior of QDs

The electrochemical behavior of QDs revealed the quantized electronic behavior as well as decomposition reactions upon reduction and oxidation. As back in 2001, Bard et al. [1] had fully investigated the direct correlation between the electrochemical bandgap and the electronic spectra of CdS QDs in N,N′-dimethylformamide (DMF). Their research revealed that CdS QDs could act as multielectron donors or acceptors at a given potential due to trapping of holes and electrons within the particle. On the other hand, the surface structure of QDs also plays a key role in determining the properties of the particles. Unpassivated surface atoms can form electronic traps for electrons and holes. Later, Bard's group studied the differential pulse voltammetric (DPV) behavior of trioctylphosphine-oxide (TOPO)-capped CdTe QDs in dichloromethane and a mixture of benzene and acetonitrile [2], as shown in Fig. 5.1.

The DPV of TOPO-capped CdTe NPs owned the bandgap of about 2.1 eV and two discrete anodic peaks which were resulted from diffusion of NPs in solution.

Fig. 5.1 DPV of two different batches of CdTe NPs at a 0.06-cm² Pt working electrode with scan toward positive or negative potentials. **a** 9.6 μM CdTe NPs in 5:1 (v/v) benzene/ acetonitrile containing 0.1 M TBAP. **b** 32 μM CdTe NPs in CH₂Cl₂ containing 0.1 M TBAPF6. Reproduced with permission from Ref. [2]. Copyright 2004, American Chemical Society

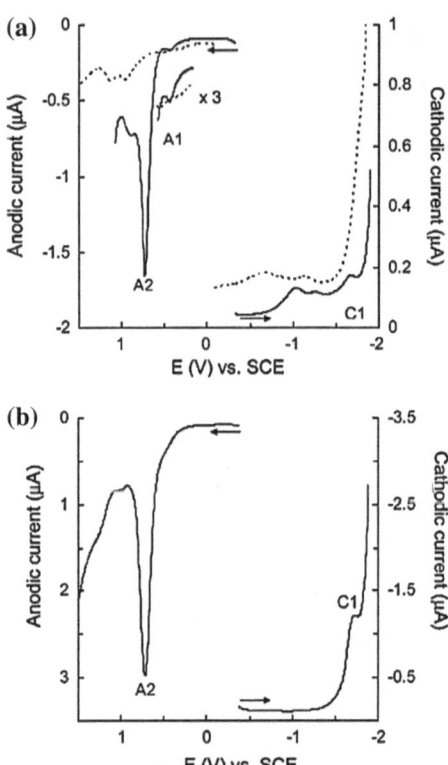

One large anodic peak at 0.7 V was proposed as a multielectron reaction, and the other anodic peak appeared because of oxidation of reduced species. Gao et al. [3] and Greene et al. [4] reported the voltammetric current peaks of QDs in aqueous solution and concluded that the electrochemical bandgap was located at potentials inside the valence-band edge, which was explained by hole injection into the sur-face traps of the particles.

To deeply understand the QDs electrochemical property, a full investigation into the effects of different parameters on the QDs electrochemical response was needed, which included the QDs size, the capping stabilizer, the value of pH, and the coexisted chemicals. Given the fact that the cyclic voltammetry (CV) is very sensitive to the nanocrystalline surface state and could provide complimentary information for a better understanding of the special size-dependent properties of semiconductor QDs, the size effect on the reduction and oxidation potentials was studied via CV in an aqueous buffer solution with the thiol-capped CdTe QDs as the object [5]. CV studies of CdTe in aqueous solution demonstrated that the size effect on the reduction and oxidation potentials could be attributed to the energetic band positions, owing to the quantum size effect. In contrast to a prediction based on the quantum size effect, the oxidation peak moves to more negative potential as the nanocrystalline size decreases.

Using the mercaptopropionic acid (MPA)-capped CdTe QDs as the object, the effects of capping stabilizer, the value of pH, and coexisted chemicals on electrochemical responses of QDs were investigated in detail via DPV [6]. Three DPV peaks (A1 at 0.36, A2 at 0.68, and A3 at 0.84 V) could be observed in the MPA-capped CdTe QDs solution, which indicated that three electrochemical processes existed in the single scan. And the conclusion could be presented as the following:

- Different stabilizers showed little effects on the existence of A1.
- The value of pH showed important effect on DPV response of the MPA-capped CdTe QDs. A1 only existed in a pH range from 5.2 to 8.0 with the maximum response at pH 6.0. A2 and A3 merged with each other simultaneously in pH 6.6 and became one peak completely with pH value higher than 7.0.
- Coexisted chemicals showed different effects on DPV response of MPA-capped CdTe QDs. Electroinactive chemicals, like chlorobenzene, showed little effect on DPV response. ECL coreactant, such as oxalic acid, hydrogen peroxide, and persulfate, also showed little effect on A1 process. Magnesium nitrate could dramatically suppress all the processes A1, A2, and A3, while potassium nitrate had little effect on A1.

Recently, Amelia et al. [7] systemically summarized the electrochemical properties of CdSe and CdTe QDs. By utilization of the most common electrochemical methods such as voltammetric methods and spectroelectrochemistry, they fully investigated the electrochemical studies of core and core–shell semiconductor nanocrystals of spherical shape. The results of representative studies were carried out taking CdSe and CdTe as examples. Different techniques by different groups were compared in order to attempt an interpretation of sometimes contradictory results.

As new type of QD, graphene QDs have been widely studied nowadays. As back in 2004, Compton and coworkers [8] have fully investigated the electrochemical characteristics of highly ordered pyrolytic graphite (HOPG) and found that the edge plane sites/defects of the HOPG were the predominant electrochemical active sites. Later, Daniel's group and Robert's group studied the electrochemical behavior of monolayer graphene sheets, respectively, whose results were both published in ACS Nano sequentially. Daniel's [9] group first performed electrochemical studies of individual monolayer graphene sheets derived from both mechanically exfoliated graphene and CVD graphene. They concluded that the electron transfer rates of graphene electrodes are more than tenfold faster than the basal plane of bulk graphite, which could be contributed to the presence of corrugations in the graphene sheets. By investigating the electrochemical properties of CVD-grown graphene electrodes in FcMeOH electrolyte at different scan rates, they found that the effective surface area of the graphene electrode was less than the geometric area of this electrode, indicating that the redox reactions occurred predominantly on a clean graphene surface. Kinetic parameter of ΔE_p (after proper resistance correction) ranged from 68.6 to 72.6 mV and increased at higher scan rate, indicative of quasi-reversible kinetics in the system, which was directly proportional to the reciprocal of the square root of scan rate, $v^{-1/2}$ (Fig. 5.2).

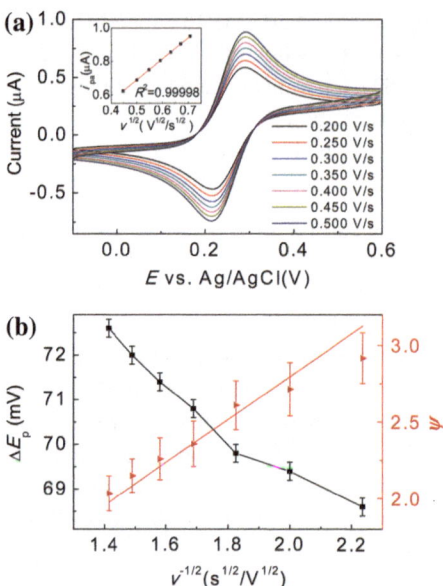

Fig. 5.2 Electrochemistry at a CVD graphene electrode. **a** Cyclic voltammograms of FcMeOH (1 mM) in $H_2O/0.1$ M KCl measured at a CVD graphene electrode at different potential scan rates. *Inset* plot of the anodic peak current (ip) versus the square root of the potential scan rate ($v^{1/2}$). **b** Peak separation ΔEp and Nicholson's kinetic parameter ψ versus the reciprocal of the square root of the potential scan rate ($v^{-1/2}$). A linear fit is used to determine the standard heterogeneous charge transfer rate constant ($k0$). Reproduced with permission from Ref. [9]. Copyright 2011, American Chemical Society

Afterward, Robert's group [10] researched the electrochemical properties of the exfoliated single and multilayer graphene flakes to measure the rate constant for electron transfer. Mechanical exfoliated graphene flakes were deposited on silicon/silicon oxide wafers to make the masked graphene/graphite samples as the working electrode. They found that both multilayer and monolayer graphene microelectrodes showed quasi-reversible behavior during voltammetric measurements in potassium ferricyanide.

In recent years, graphene as a new class of two-dimensional nanomaterial has attracted considerable attention. The excellent electronic transfer rate, single-layered structure, and good biocompatibility endow graphene with great potential applications in the field of electrocatalytic biodevices [11–13]. Wang's group [14] reported on the utilization of graphene–CdS nanocomposite as a novel immobilization matrix for the GOD immobilization. The nanocomposite could provide a unique microenvironment for the direct electrochemistry of GOD, and the immobilized GOD on the modified electrode possessed its native structure and electrocatalytic activities. In comparison with the graphene sheets and CdS nanocrystals, the graphene–CdS nanocomposites exhibited excellent electron transfer properties for GOD with a rate constant (k_s) of 5.9 s^{-1} due to the synergy effect of graphene sheets and CdS nanocrystals. Based on the electrocatalytic response of the reduced form of GOD to dissolved oxygen, the obtained glucose biosensor displays satisfactory analytical performance over an acceptable linear range from 2.0 to 16 mM with a detection limit of 0.7 mM.

In our group, a series of graphene QD (GQD) and its nanocomposites were synthesized and used in electrochemical applications. Different from the route of traditional nanolithography and the chemical breakdown of graphene oxide (GO), we produced GQDs with different size distribution in scalable amounts with acidic exfoliation and etching of carbon fibers [15]. The stacked graphitic submicrometer domains of the fibers could be easily broken down during the acid treatment and chemical exfoliation of traditional pitch-based carbon fibers. The size of the as-prepared GQDs varies with the reaction temperature, and the emission color and the bandgap of GQDs can be controlled accordingly. By utilization of its good biocompatibility, high water solubility, and low toxicity, GQD and its composite were employed to fabricate kinds of biosensors. In 2012, graphene–CdS (GR–CdS) nanocomposites were prepared in a one-step synthesis in aqueous solution. GO was simultaneously reduced to GR during the deposition of CdS. The heteronanostructure of the as-prepared GR–CdS nanocomposite films could facilitate the spatial separation of the charge carriers, which endows nanomaterial with the excellent electrontransport properties. When used for the fabrication of an advanced photoelectrochemical cytosensor, the GR–CdS nanocomposite-based biosensor showed a good photoelectronic effect and cell-capture ability and had a wide linear range and low detection limit for HeLa cells [16]. Later, another composite of anatase TiO_2–graphene (ATG) nanocomposites was synthesized via a one-step approach using titanium(III) ion as reductant and titanium source in an aqueous solution [17]. The high surface area, excellent conductivity, and sufficiently functional groups enable the ATG nanocomposites to be favorable for fabricating biosensors. When used for hemoglobin (Hb) immobilization, it could realize the enhanced direct electron transfer (DET) of Hb, and Hb–ATG nanohybrid exhibited good electrocatalytic activity toward the reduction of H_2O_2.

5.1.2   The Electrochemical DNA Analysis of QDs

DNA analysis is associated tightly with tissue matching, genetic diseases, and forensic applications in molecular diagnostics [18, 19]. Sensitive detection of specific nucleic acid sequences on the basis of the hybridization reaction is the key point for various applications including clinical diagnosis, environmental control, and forensic analysis [20]. Given the fact that the QDs own the excellent biocompatibility, they play an important role in DNA analysis [21].

In order to fully investigate the interaction between the QDs and the DNA, an electroactive dsDNA indicator of $Co(phen)_3^{3+/2+}$ (phen = 1,10-phenanthroline) was used to measure the dissociation behavior of double-stranded DNA (dsDNA) via the electrochemical technique [22]. It was found that $Co(phen)_3^{3+/2+}$ was more easily dissociated from dsDNA-modified gold electrode in the presence of CdTe QDs. In relatively low ionic strength, the dissociation coefficient constant of $Co(phen)_3^{3+/2+}$ in the presence of CdTe QDs was 3.1 times higher than that in the absence of CdTe QDs. This value reduced to 1.32 times in relatively high ionic strength. This indicated that the binding site of CdTe QDs on dsDNA was probably at major groove of dsDNA. This demonstration offers a new approach to illustrate the QDs cytotoxicity mechanism. Based on this research, Jiao et al. [23] developed an electrochemical biosensing for dsDNA damage induced by PbSe QDs under UV irradiation. In this research, the damage of dsDNA was fulfilled by immersing the sensing membrane electrode in PbSe QDs suspension and illuminating it with an UV lamp. Cyclic voltammetry was utilized to detect dsDNA damage with $Co(phen)_3^{3+}$ as the electroactive probe. The synergistic effect among the UV irradiation, Pb^{2+} ions liberated from the PbSe QDs under the UV irradiation, and the reactive oxygen species (ROS) generated in the presence of the PbSe QDs dramatically enhanced the damage of dsDNA. This electrochemical sensor provided a simple method for detecting DNA damage and may be used for investigating the DNA damage induced by other QDs.

Using CdSe/ZnS as label, a relatively simple, time-saving, and multiapproach biosensor for the DNA detection was fabricated in our group [24]. By detecting the cadmium content in the bond QDs, the target DNA could be indirectly detected through the SWASV assay. Based on the hairpin probe and site-specific DNA cleavage of restriction endonuclease, Chen et al. [25] fabricated an electrochemical DNA biosensor. This biosensor was used to detect DNA species related to cymbidium mosaic virus. The stripping voltammetric measurements of the dissolved Cd^{2+} were successfully performed to indirectly determine the sequence-selective discrimination between perfectly matched and mismatched target DNAs including a single-base mismatched target DNA, and the limit detection could reach as low as 3.3×10^{-14} M for complementary target DNA. Given the simplicity in design of the proposed electrochemical sensor, it is fairly easy to generalize this strategy to detect a spectrum of targets and might have a promising future for the investigation of DNA hybridization, also would play the potential predominance in diagnosis of virus or diseases.

What is more, based on the fact that different metal components of different QD nanocrystal tracers yield different well-resolved and highly sensitive stripping voltammetric signals, the multitarget electrochemical biosensor could be fabricated via the utilization of different QD codes. This new multielectrochemical coding technology opens new opportunities for DNA diagnostics and for bioanalysis.

In 2003, Wang et al. [26] first employed this strategy for the simultaneous detection of multiple DNA targets based on QD tags with diverse redox potentials (Fig. 5.3). Such encoding QDs offered a voltammetric signature with distinct electrical hybridization signals for the corresponding DNA targets. Via the utilization of different inorganic colloid QD nanocrystal tracers, whose metal components yield well-resolved highly sensitive stripping voltammetric signals for the corresponding targets, three encoding QDs (ZnS, CdS, and PbS) have thus been used

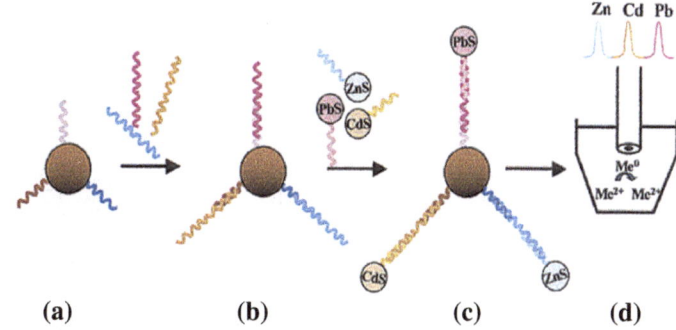

Fig. 5.3 Multitarget electrical DNA detection protocol based on different inorganic colloid nanocrystal tracers. **a** Introduction of probe-modified magnetic beads. **b** Hybridization with the DNA targets. **c** Second hybridization with the QD-labeled probes. **d** Dissolution of QDs and electrochemical detection. Reproduced with permission from Ref. [26]. Copyright 2003, American Chemical Society

to differentiate the signals of three DNA targets in connection with a sandwich hybridization assay and stripping voltammetry of the corresponding heavy metals. The new multitarget electrical detection scheme incorporates the high sensitivity and selectivity advantages of QD-based electrical assays.

5.1.3 The Electrochemical Aptamer Analysis of QDs

As a new class of single-stranded DNA/RNA molecules, aptamers have received a great deal of attention and attracted much interest in recent years. Aptamers are selected from synthetic nucleic acid libraries via the selection procedure called systematic evolution of ligands by exponential enrichment (SELEX) [27, 28]. Aptamer owns the ability to form defined tertiary structures upon specific target binding. Since its first discovery in 1990s, many aptamers have been selected for corresponding target combination ranging from metal ions, organic molecules, biomolecules, to entire organism and even whole cells [29–33]. Compared with natural receptors such as antibodies and enzymes, aptamers could be simply and reproducibly synthesized and easily labeled [34, 35]. Besides, aptamers have high flexibility and could be modified with certain functional groups in biosensor design. Because of these important features, more and more interests are attracted in developing aptamer-based biosensor (aptasensor) [36–40].

With QDs coupled with different analytes, different targets could be detected via DPV and SWASV by fabricating the aptasensor, such as ATP [41], thrombin [42], cocaine [43], etc. In our group, the three-dimensionally ordered macroporous (3DOM) gold film was used, instead of the classical bare flat Au electrode, to fabricate a sensitive electrochemical aptasensor for the detection of ATP [41],

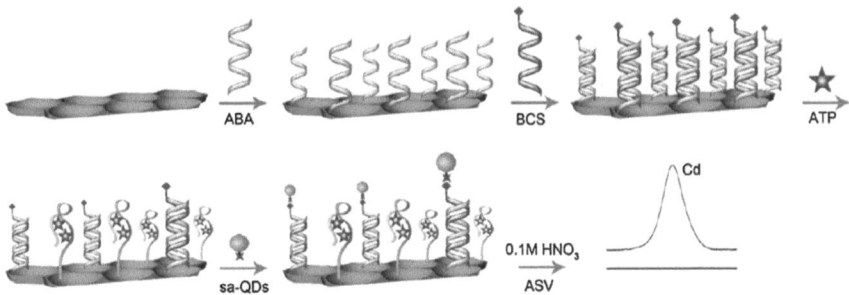

Fig. 5.4 Schematic illustration of the QDs (CdSe/ZnS) electrochemical aptasensor based on 3DOM gold film. Reproduced with permission from Ref. [41]. Copyright 2010, Elsevier

as shown in Fig. 5.4. The 3DOM gold film endowed the active surface area of the electrode up to 9.52 times larger than that of a classical bare flat one. The reaction was monitored by electrochemical stripping analysis of dissolved QDs which were bound to the residual cDNA through biotin–streptavidin system. The decrease in peak current was proportional to the amount of ATP. The unique interconnected structure in 3DOM gold film along with the "built-in" preconcentration remarkably improved the sensitivity down to 0.01 nM. This promised a novel model for the detection of small molecules with higher sensitivity.

The multicomponent analysis could not only be used in the DNA analysis, but also be employed in the field of aptasensor. As back in 2006, based on their multielectrochemical coding technology for the simultaneous detection of multiple DNA targets, Wang and cooperators described a simple method for preparing a QD/aptamer-based ultrasensitive multianalyte electrochemical biosensor with subpicomolar (attomole) detection limits [44]. The main strategy is accomplished using a simple single-step displacement assay, which involved the coimmobilization of several thiolated aptamers, along with binding of the corresponding QD-tagged proteins on a gold surface. After the addition of the protein sample without QD label, monitoring the displacement through electrochemical detection of the remaining QDs could be achieved. Such electronic transduction of aptamer–protein interactions is extremely attractive for meeting the low power, size, and cost requirements of decentralized diagnostic systems. What is more, unlike two-step sandwich assays, the new aptamer biosensor protocol relies on a single-step displacement protocol.

Most recently, Yuan's group [43] developed a "signal on" and sensitive biosensor for one-spot simultaneous detection of multiple small molecular analytes based on electrochemically encoded barcode QD tags. In this route, the target analytes of adenosine triphosphate (ATP) and cocaine are respectively sandwiched between the corresponding set of surface-immobilized primary binding aptamers and the secondary binding aptamer/QD bioconjugates. The captured QDs (CdS and PbS) yield distinct electrochemical signals after acid dissolution. Due to the inherent amplification feature of the QD labels and the "signal on" detection scheme, as well as the sensitive monitoring of the metal ions released upon acid dissolution of the QD labels, low detection limits of 30 and 50 nM were obtained for ATP and cocaine, respectively, in these assays. The high specificity to target analytes and promising applicability to complex sample matrix made the proposed assay protocol an attractive route for screening of small molecules in clinical diagnosis.

5.2 QDs for Photoelectrochemical Analysis

Rapid, specific detection of nucleic acid sequences has attracted significant attention due to possible applications in fields ranging from pathogen detection to the diagnosis of genetic diseases. Among various detection techniques, photoelectrochemical detection has attracted significant interest. Firstly, this method is very sensitive with low background signals due to the different forms of energy for excitation (light) and detection (current) [45–47]. Secondly, compared to optical detection methods such as fluorescence, chemiluminescence (CL), and ECL, which use complex and expensive optical imaging devices, the instrument of photoelectrochemistry is much simpler and of low cost [48]. QDs, with their unique fluorescence properties and photoelectrochemical functions, are photoactive materials for the development of nucleic acid sensor systems. A competitive DNA hybridization assay based on the photoelectrochemistry of the semiconductor quantum dot-single stranded DNA conjugates (QD-ssDNA) was developed by Deniz et al. [49]. A sensing surface is constructed by a self-assembled monolayer (SAM) formation on ITO surface, activation of surface for immobilization of the amine-modified ssDNA (probe), and then immobilization of probe on activated surface. After obtaining sensing surface, competitive DNA assay was performed on the probe-immobilized surface and the target concentration in the sample was determined based on photocurrent measurements. As seen in Fig. 5.5, a current change, photocurrent, was observed in anodic direction when the light source was turned on, and the system immediately turned to its initial state after the light source was turned off. Upon the competition between QD-ssDNA and single-stranded target DNA, the photocurrent response decreased with the increase in the target DNA concentration. A linear relationship between the photocurrent and the target DNA concentration was obtained ($R^2 = 0.991$), and limit of detection was found to be as 2.2 μM target ssDNA. The selectivity of system toward the target DNA was also demonstrated using noncomplementary sample.

For better sensitivity, the amplified detection of DNA was accomplished by developing a novel architecture of double-stranded DNA-cross-linked CdS nanoparticle arrays on electrode supports and the structurally controlled generation of photocurrents upon irradiation of these arrays [50]. The electrostatic binding of $[Ru(NH_3)_6]^{3+}$ to the dsDNA units provides tunneling routes for the conduction-band electrons and thus results in enhanced photocurrents. CdS nanoparticles (2.6 $\pm$ 0.4 nm) were functionalized with thiolated oligonucleotide 1 or 2. These two oligonucleotides are complementary to the 5'- and 3'-ends of the target

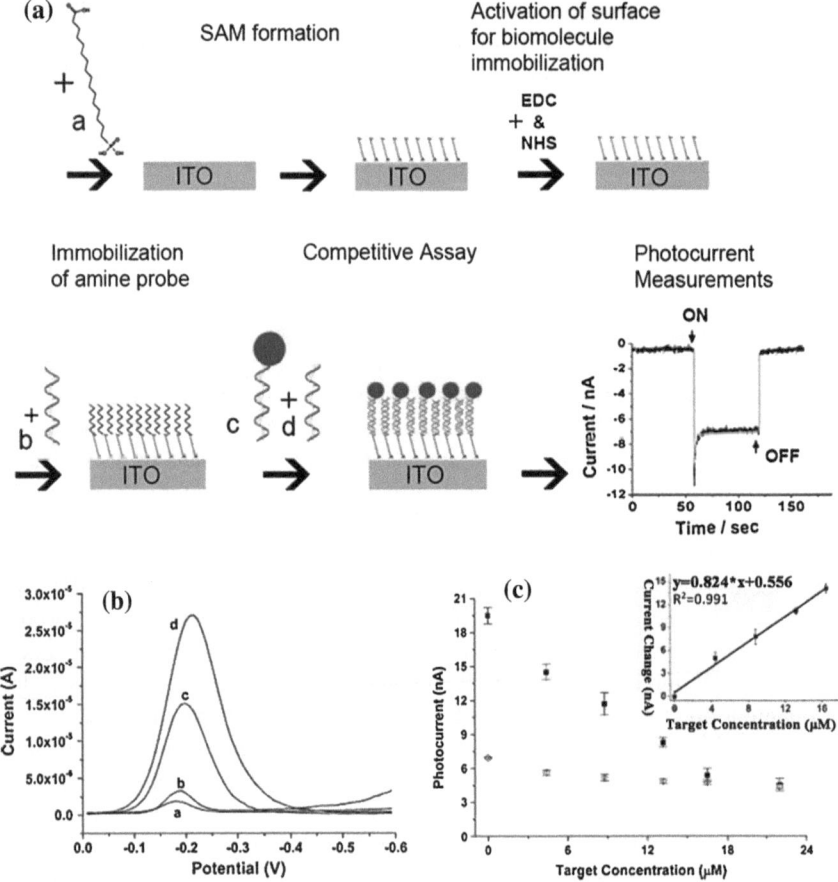

Fig. 5.5 a Schematic representation of self-assembled monolayer formation, surface activation, probe immobilization, and the competitive DNA assay. *a* 16-phosphonohexadecanoic acid, *b* amine-modified ssDNA (probe), *c* QD–ssDNA conjugate and d target ssDNA (ON and OFF refers to states of the light source). **b** Differential pulse voltammetry of MB at different electrodes: *a* SAM/ITO, *b* bare ITO, *c* after hybridization on DNA-modified ITO electrode, and *d* before hybridization on DNA-modified ITO electrode. **c** Photocurrent values for the competitive hybridization assays performed at 25 °C (*open diamond*) and 50 °C (*filled square*). Inset linear dependence of current change on the target ssDNA at 50 °C. The error bars refer to standard deviation of repetitive measurements (*n* = 3). Reproduced with permission from Ref. [49]. Copyright 2011, Springer

DNA 3. Figure 5.6 shows the stepwise assembly of the DNA-cross-linked CdS particles on an Au electrode. The oligonucleotide 1 was assembled on the Au electrode and then treated with the analyte 3 to yield the dsDNA system. Subsequent interaction of the surface with the 2-functionalized CdS resulted in the binding of the CdS nanoparticles to the surface. Further alternating interaction of the interface with a solution containing the 1-functionalized CdS nanoparticles pretreated

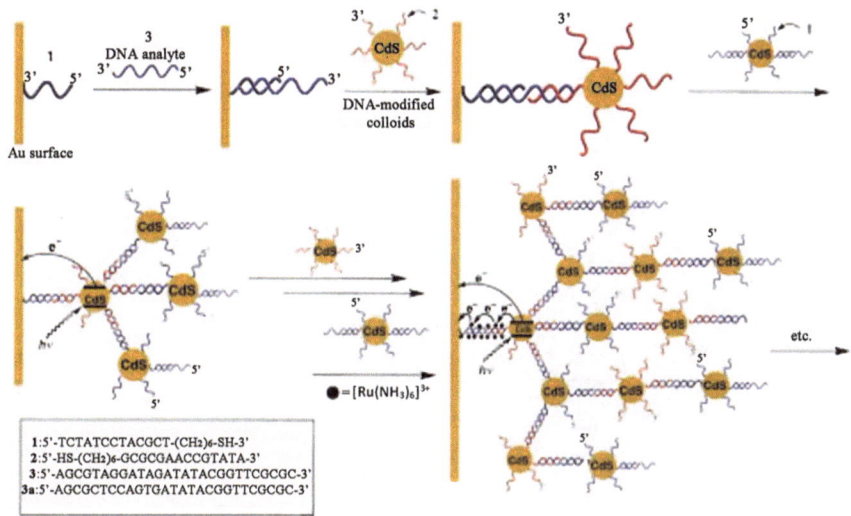

Fig. 5.6 The organization of oligonucleotide/DNA-cross-linked arrays of CdS nanoparticles and the photoelectrochemical response of the nanoarchitectures. Reproduced with permission from Ref [50]. Copyright 2001, Wiley

with 3 and with a solution containing 2-functionalized CdS nanoparticles resulted in an array with a controlled number of CdS nanoparticle generations. The photocurrent follows the absorbance spectrum of the CdS nanoparticles, and it increases with increasing number of generations of cross-linked particles. As we use a sacrificial electron donor as hole scavenger, we attribute the resulting photocurrent to the injection of conduction-band electrons into the electrode. The photocurrent can be switched "on" and "off" by pulsed irradiation of the arrays. The mechanism of photocurrent generation probably involves the photoejection of conduction-band electrons of CdS particles in contact with or at tunneling distances from the electrode.

The conductivity of DNA has been a subject of extensive controversy [51], and it is accepted that DNA exhibits poor conductivity [52]. However, the conductivity of DNA could be controlled by appropriate ordering of the base sequence [53] or by the incorporation of redoxactive intercalators into dsDNA [54]. CdS nanoparticle/DNA conjugates was immobilized on gold surfaces and the effects of intercalators and the applied potential on the photoelectrochemical features of the system was described in Fig. 5.7 [55]. The results demonstrate that the resulting photocurrent can be reversibly switched between cathodic and anodic directions by controlling the redox state of the intercalated species. The intercalation of doxorubicin into the dsDNA here results in a fivefold higher anodic photocurrent. The intercalation of methylene blue into the dsDNA results in enhanced cathodic photocurrent while the intercalation of oxided methylene blue in enhanced anodic photocurrent.

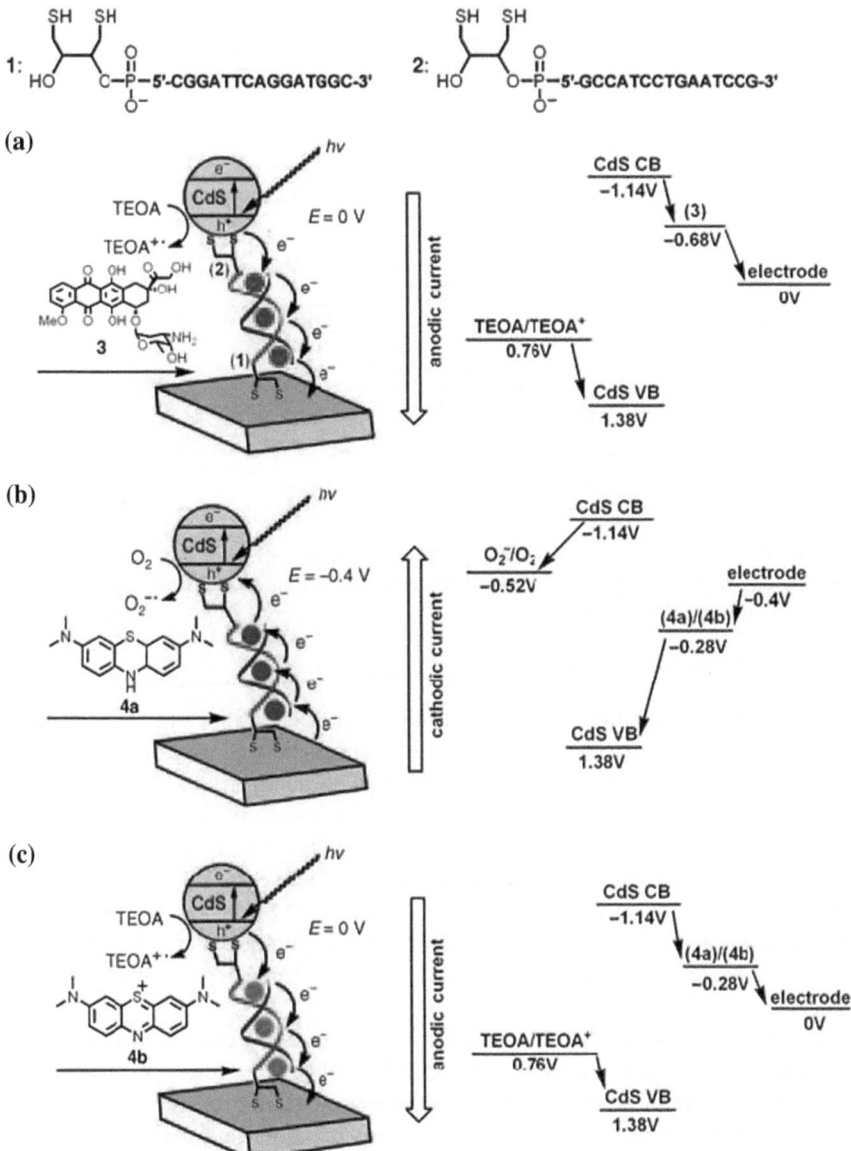

Fig. 5.7 Directional electroswitchable photocurrents in the CdS nanoparticle/dsDNA/intercalator system. **a** Enhanced generation of an anodic photocurrent in the presence of the doxorubicin intercalator (3; applied potential $E = 0$ V). **b** Enhanced generation of a cathodic photocurrent in the presence of the reduced methylene blue intercalator (4a; applied potential $E = 0.4$ V). **c** Enhanced generation of an anodic photocurrent in the presence of the oxidized methylene blue intercalator (4b; applied potential $E = 0$ V). The redox levels of the components that participate in the different photocurrent-generating systems are presented on the right side of the scheme. *TEOA* triethanolamine, *CB* conduction band, *VB* valence band. Reproduced with permission from Ref. [55]. Copyright 2005, Wiley

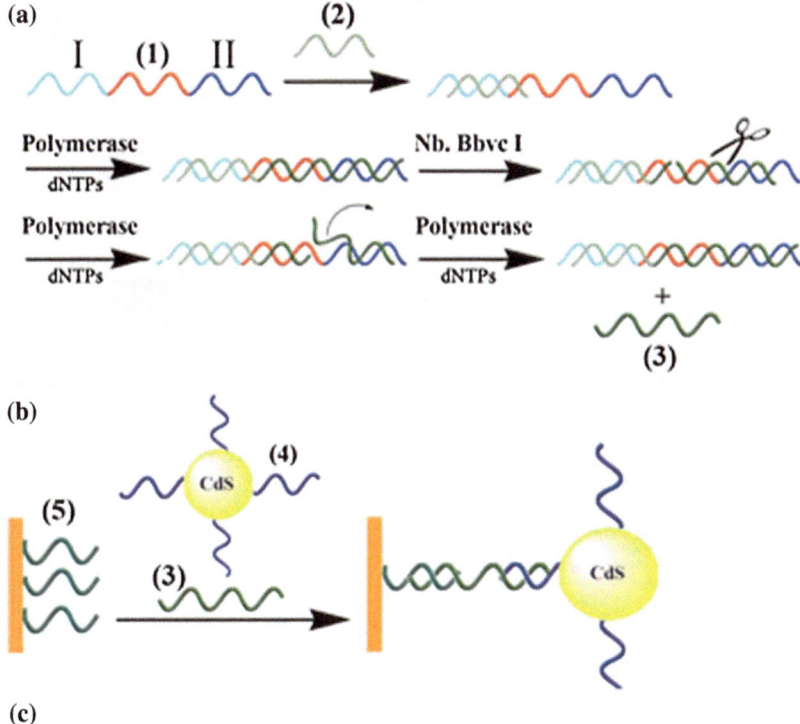

(c)

1) GCGCGAACCGTATATCTATCCTACGCTCCTCAGCCCACACGATCCT
2) AGGATCGTGTG
3) TGAGGAGCGTAGGATAGATATACGGTTCGCGC
4) H₂N-(CH₂)₆-GCGCGAACCGTATA
5) TCTATCCTACGCT-(CH₂)₆-SH

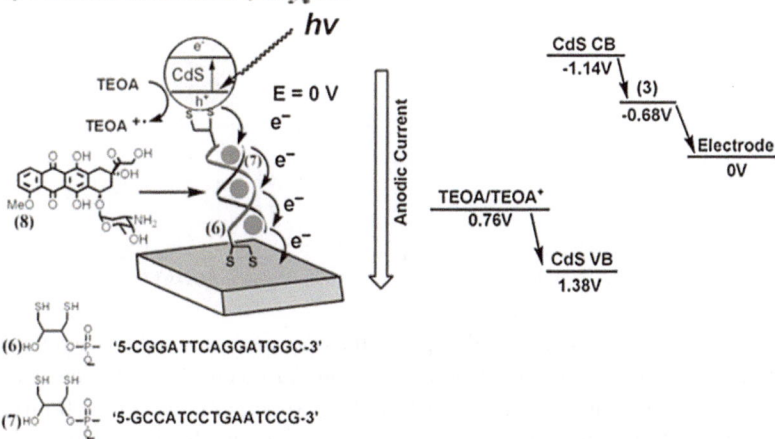

Fig. 5.8   a The autonomous synthesis of the nucleic acid "waste product" by the DNA machine;
b assembly of the CdS NPs on the nucleic acid-functionalized Au electrode using (3) as glue
units. **c** Assembly and photoelectrochemical functions of a CdS NP-functionalized duplex DNA
associated with a Au electrode. Reproduced with permission from Ref. [56]. Copyright 2007,
Royal Society of Chemistry

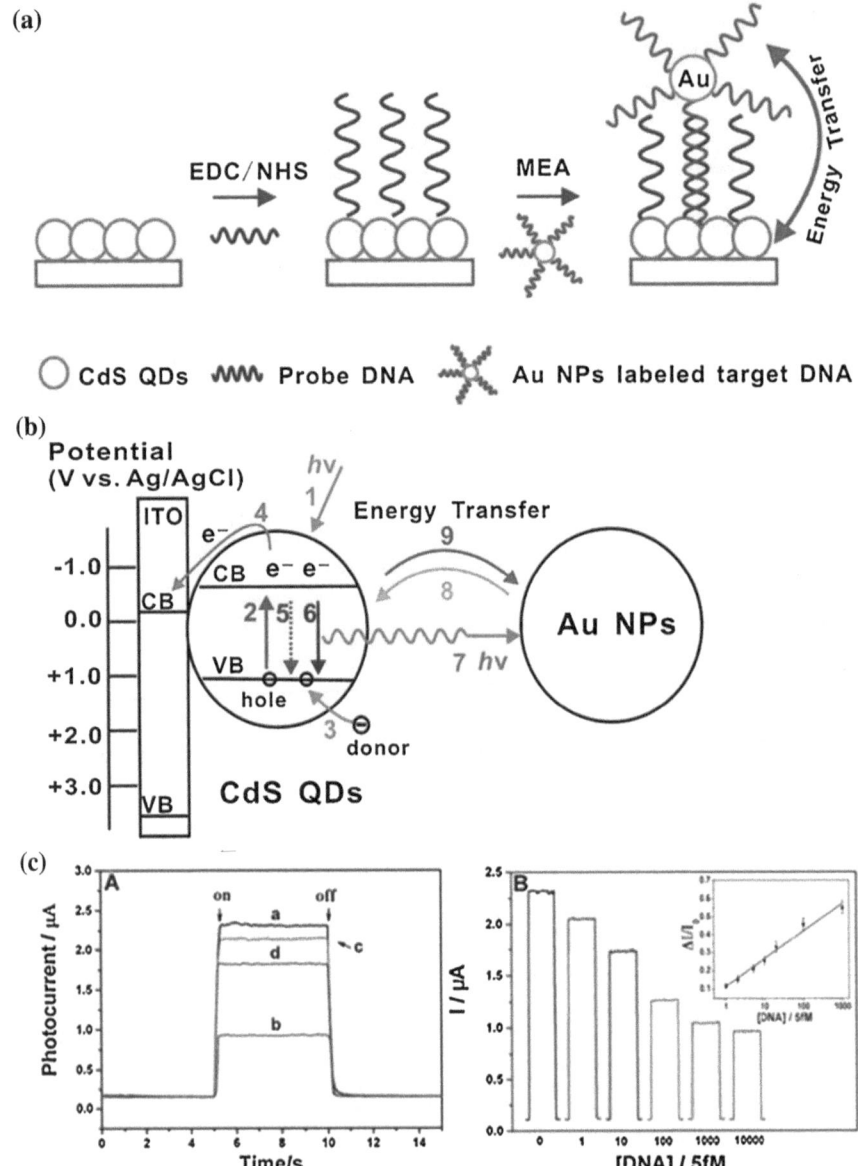

This basic phenomenon is used to probe the operation of a DNA-based machine through the assembly of CdS nanoparticles on Au electrode [56]. The machine includes a nucleic acid "track," (1) that binds a primer, (2), through hybridization to a predefined domain. Figure 5.8 depicts the principles for the DNA-machine-induced assembly of aggregated CdS nanoparticles on an electrode support. The synthesis of the DNA occurs on a "DNA track" (1) that consists of two regions I and II, Fig. 5.8a. Region I is complementary to the primer (2). The sequence II is

Fig. 5.9 **a** Schematic diagram of the process to fabricate the energy transfer-controlled PEC system. **b** Schematic mechanism of the operating PEC system. Process 1, photoexcitation of the CdS QDs; 2, photon absorption and electron transfer from the valence band (*VB*) to the conduction band (*CB*); 3, hole neutralization by electron donor; 4, electron ejection to the electrode for photocurrent generation; 5, nonradiative electron–hole recombination; 6, radiative electron–hole recombination; 7, spontaneous emission originating from radiative decay; 8, plasmon enhancement on radiative decay; and 9, exciton energy transfer (*EET*) from CdS QDs to Au NPs. **c** Photocurrent intensity in 0.10 M PBS containing 0.08 M ascorbic acid of (*a*) CdS/ITO electrode modified with 20 mL, 1 mM capture DNA and blocked by MEA, and after hybridization with (*b*) Au NP-labeled target DNA, (*c*) bare target DNA, and (*d*) SiO$_2$ NP-labeled target DNA. **d** Effect of different concentrations of target DNA on the differential photocurrent responses. *Inset* the corresponding calibration plot ($\Delta I = I_0 - I$, I_0 and I are the photocurrents of the capture DNA/CdS/ITO electrode prior to and after hybridization). The working potential was 0.0 V, and the excitation wavelength was 420 nm. Reproduced with permission from Ref. [57]. Copyright 2011, Royal Society of Chemistry

complementary to a nucleic acid (3) that is being released by the machine. Upon the hybridization of (2) with the region (I) of (1) and the subsequent addition of polymerase, the dNTPs mixture, and the nicking enzyme Nb. BbvCI, polymerization of the complementary strand to (1) is initiated. The replication of domain II yields, however, the key for the machine operation since the generated strand in the duplex of region II includes the nicking site for the enzyme. Thus, while the initial replication of the template proceeds, the nicking process establishes a new polymerization site. The subsequent polymerization involves the displacement of the originally replicated strand (3). Thus, the replication, nicking, and displacement of (3) proceed autonomously, and the displaced product (3) may be viewed as a "waste product." The "waste product" (3) is then used as the bridging unit for the assembly of the CdS nanoparticles. CdS nanoparticles were modified with the thiolated nucleic acid (4) that is complementary to the 5-end of (3) Fig. 5.8b. The Au electrode was functionalized with the thiolated nucleic acid, (5), that is complementary to the 3-end of (3). The initiation of the machine operation in the presence of the (5)-modified electrode results in the hybridization of (3) with the electrode and the subsequent hybridization of the (4)-modified CdS nanoparticles with the modified electrode, Fig. 5.8b. Thus, the photoelectrochemical response of the machine may act as a transduction signal for the time-dependent assembly of the CdS nanoparticle aggregates.

The above works in this field are exclusively based on the change in direct electron transfer process between the photoactive materials and the ambient environment prior to and after the biorecognition events. Chen et al. present the first exploitation of energy transfer between CdS QDs and Au NPs in a PEC system to search for an advanced energy transfer-based PEC bioassay protocol [57]. As shown in Fig. 5.9a, the new energy transfer-based PEC detection format involved the modification of an indium tin oxide (ITO) electrode with CdS QDs, followed by the integration of Au NPs into the system for sensitive DNA detection, on the basis of the interparticle energy transfer between CdS QDs and Au NPs bridged by the biorecognition of DNA. As shown in Fig. 5.9b, photoexcitation of the CdS

QDs (process 1) would result in the electron transfer from the valence band (VB) to the conduction band (CB) (process 2), thus yielding electron–hole pairs. As soon as the charge separation occurs, the electron–hole pairs would be destined for recombination (process 5 or 6) or the charge transfer (such as process 3 and 4). The emission of the CdS QDs was used to excite the surface plasmon resonance (SPR) of the proximal Au NPs (process 7), which would create local electric fields that in turn modulate the exciton states in the CdS QDs by enhancing the radiative decay rate (process 8). Along with the SPR effect, Au NPs would also introduce an additional nonradiative decay route for electron–hole recombination in CdS QDs (process 5) with exciton energy transfer (EET) from the CdS QDs to the Au NPs (process 9). Processes 5 and 6 are cooperative, and their overall effect contends with the electron transfer of process 4, so the concern here is their overall effect on the excitonic response of the CdS QDs, and hence, on the final photocurrent intensities, that could facilely be monitored by electrical signal. The photocurrent decrease was proportional to the DNA concentration logarithmically with the linear range from 5.0×10^{-15} M to 5.0×10^{-12} M ($R^2 = 0.9858$) and detection limit of 2.0×10^{-15} M. Comparing with gold, silver has a dielectric function, $\varepsilon_{Ag}(\omega)$, and a stronger plasmon resonance. Plasmon band of Ag NPs fully overlaps with the absorption band of utilized CdS QDs Ag-NPs-based assemblies can demonstrate enhanced properties suitable for optical and sensor applications [58]. Due to their natural absorption overlap, the exciton of the QDs and the plasmon of Ag NPs could be induced simultaneously. The EPI resonant nature enabled manipulating photoresponse of the QDs via tuning interparticle distances (Fig. 5.10). Specifically, the photocurrent of the QDs could be greatly attenuated and even be completely damped by the generated EPI. The photocurrent decrease was proportional to the concentration of labeled target DNA in logarithmic scale with the linear range from 2.0×10^{-15} to 2.0×10^{-11} M.

In order to produce photocurrents, photoexcitation of light source is usually required. However, the appendant light source makes the instrument complicated. In addition, the exciting wavelengths of various photoelectroactive materials are different. Hence, monochromator is needed to bring appropriate exciting light, which makes the volume of the instrument bigger and departures from the portable trend for biosensor. Consequently, a strategy for substitution of physical light source is highly deserved [59]. Chemiluminescence is defined as a process in which excited molecules or atoms generated from chemical reactions release the excess of energy in light form. Different CL systems can bring emission light of various wavelengths. At the same time, reaction conditions such as type of fluorescence reagent and reaction solution can also affect the emission wavelength. Thus, by adjusting the conditions of CL reaction, various photoelectrochemically active species can be excited theoretically, which can realize the photoelectrochemical detection free from physical light source. The photoelectrochemical analysis of a DNA analyte without an external irradiation of the QD-modified electrode is depicted in Fig. 5.11a. The CdS QDs linked to the electrode were functionalized with BSA units, and the thiolated nucleic acid probe, 3, that is complementary with the 5'-end of the analyte, 4, was tethered to the BSA layer. The measurement

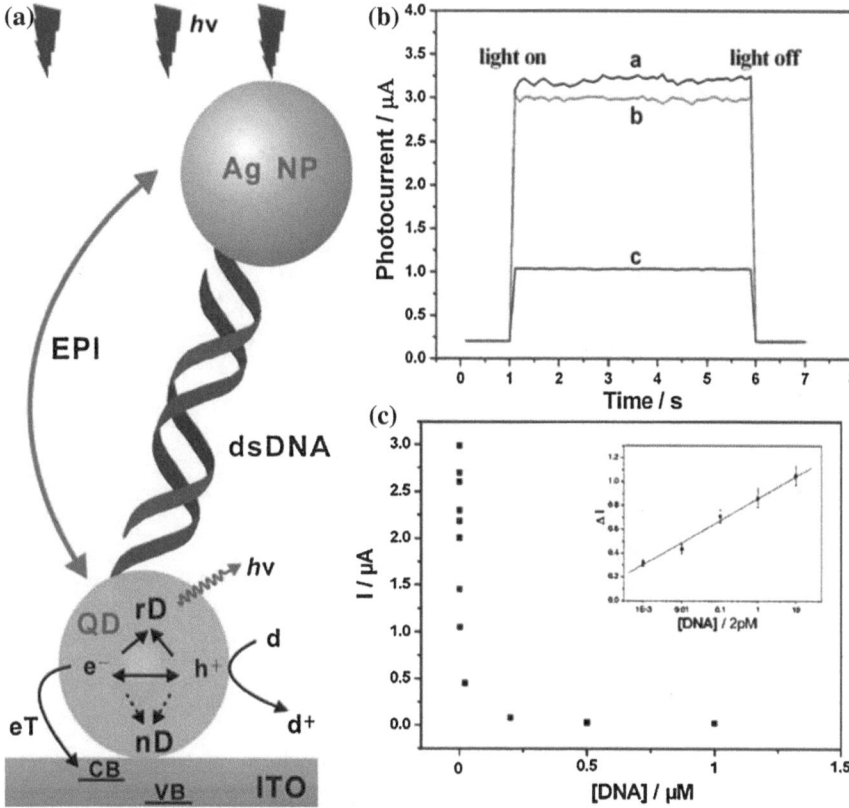

Fig. 5.10 a Schematic mechanism of the operating PEC system. **b** Photocurrent intensity in 0.10 M PBS containing 0.10 M ascorbic acid of (*a*) PDDA/CdS-modified ITO electrode, (*b*) modified with 20 µL, 1 µM capture DNA and blocked by MEA, and (*c*) hybridized with Ag NPs-labeled target DNA. The oligonucleotide sequences of 12 base pairs were used here, and their concentration was 1.0×10^{-6} M. The working potential was 0.0 V, and the excitation wavelength was 420 nm. **c** Effect of different concentrations of target DNA on the differential photocurrent responses ($\Delta I = I0-I$, $I0$ and I are the photocurrents of capture DNA/CdS/ITO electrode before and after hybridization). *Insert* the corresponding calibration curve. The used DNA was 36 bases, and the photocurrent measurement was carried out in 0.10 M PBS containing 0.10 M ascorbic acid (AA). The working potential was 0.0 V, and the light wavelength was 420 nm. Reproduced with permission from Ref. [58]. Copyright 2012, American Chemical Society

solution also contained the catalytic DNA label, **5**, composed of the G-quadruplex sequence and a nucleic acid residue that can hybridize with the free 3′-end of the analyte, **4**. Thus, the analyte serves as a bridging unit for the assembly of the hemin/G-quadruplex label on the electrode. Accordingly, in the presence of the analyte, the hemin/G-quadruplex catalytic label is positioned in close proximity to the CdS QDs associated with the electrode. As a result, the hemin/G-quadruplex-catalyzed oxidation of luminol by H_2O_2 and the resulting chemiluminescence may stimulate the CRET process and lead to the generation of the photocurrent.

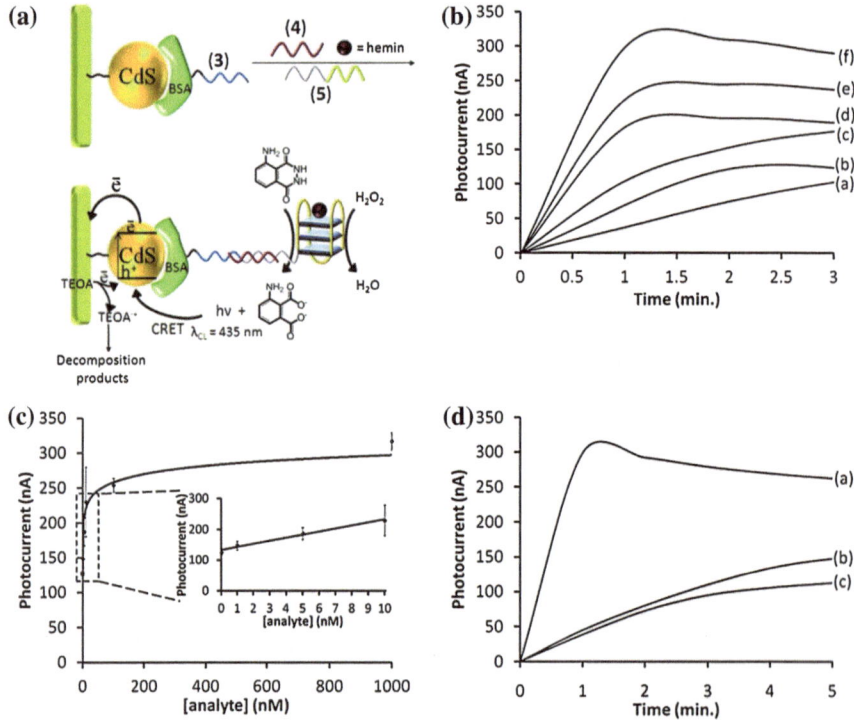

Fig. 5.11 **a** Schematic analysis of a target DNA by a sandwich-type nucleic acid assay on the CdS QDs–BSA conjugate associated with an electrode using the hemin/G-quadruplex as a catalytic label for the CRET-stimulated generation of photocurrents. **b** Time-dependent photocurrent intensities upon analyzing different concentrations of the analyte DNA, 4, using the CdS QDs/BSA-3 as the sensing interface and the hemin/G-quadrplex as a catalytic label for the generation of chemiluminescence and CRET-stimulated generation of photocurrents. Concentrations of 4 correspond to (*a*) 0 nM, (*b*) 1 nM, (*c*) 5 nM, (*d*) 10 nM, (*e*) 100 nM, and (*f*) 1,000 nM. **c** Derived calibration curve corresponding to the photocurrent intensities generated by different concentration of 4 after a time interval of 1 min. Error bars were derived from a set of $N = 3$ experiments. Inset: the linear part of the calibration curve. **d** Photocurrent intensities upon analyzing (*a*) the target, 4, 1 μM; (*b*) the single-base mismatched nucleic acid 4a, 1 μM; and (*c*) the two-base mismatched nucleic acid 4b, 1 μM, using CdS QD–BSA-3 sensing interface and 5 as a catalytic label. In all experiments, a 10 mM HEPES buffer solution, pH = 9.0, that included 20 mM KNO_3, 200 mM $NaNO_3$, 1 μM hemin, 0.1 μM 5, 0.57 mM luminol, 0.5 mM H_2O_2, and 20 mM triethanolamine (TEOA) was used as the analysis reaction medium. Reproduced with permission from Ref. [59]. Copyright 2012, American Chemical Society

Since the surface coverage of the hemin/G-quadruplex on the electrode is controlled by the concentration of the analyte, the resulting photocurrent intensities depend on the concentration of the analyte and provide a quantitative measure for the DNA analyte, 4. Figure 5.11b shows the photocurrent intensities generated by the system in the presence of different concentrations of the analyte DNA. Figure 5.11c shows the resulting calibration curve, implying that the analyte DNA could be detected with a detection limit that corresponds to 2 nM.

References

1. Haram SK, Quinn BM, Bard AJ (2001) Electrochemistry of CdS nanoparticles: a correlation between optical and electrochemical band gaps. J Am Chem Soc 123:8860–8861
2. Bae Y, Myung N, Bard AJ (2004) Electrochemistry and electrogenerated chemiluminescence of CdTe nanoparticles. Nano Lett 4:1153–1161
3. Gao M, Sun J, Dulkeith E, Gaponik N, Lemmer U, Feldmann J (2002) Lateral patterning of CdTe nanocrystal films by the electric field directed layer-by-layer assembly method. Langmuir 18:4098–4102
4. Greene IA, Wu F, Zhang JZ, Chen S (2003) Electronic conductivity of semiconductor nanoparticle monolayers at the Air|Water interface. J Phys Chem B 107:5733–5739
5. Poznyak SK, Osipovich NP, Shavel A, Talapin DV, Gao MY, Eychmuller A, Gaponik NN (2005) Size-dependent electrochemical behavior of thiol-capped CdTe nanocrystals in aqueous solution. J Phys Chem B 109:1094–1100
6. Li J, Zou GZ, Hu XF, Zhang XL (2009) Electrochemistry of thiol-capped CdTe quantum dots and its sensing application. J Electroanaly Chem 625:88–91
7. Matteo A, Christophe L, Serena S, Alberto C (2012) Electrochemical properties of CdSe and CdTe quantum dots. Chem Soc Rev 41:5728–5743
8. Davies TJ, Moore RR, Banks CE, Compton RG (2004) The cyclic voltammetric response of electrochemically heterogeneous surfaces. J Electroanal Chem 574:123–152
9. Wan L, Cen T, Michael AL, Héctor DA, Daniel CR (2011) Electrochemistry of individual monolayer graphene sheets. ACS Nano 5:2264–2270
10. Anna TV, Ian AK, Kostya SN, Cinzia C, Axel E, Ernie WH, Robert AWD (2011) Electrochemical behavior of monolayer and bilayer graphene. ACS Nano 5:8809–8815
11. Zuo XL, He SJ, Li D, Peng C, Huang Q, Song SP, Fan CH (2010) Graphene oxide-facilitated electron transfer of metalloproteins at electrode surfaces. Langmuir 26:1936–1939
12. Shao YY, Wang J, Wu H, Liu J, Aksay IA, Lin YH (2010) Graphene based electrochemical sensors and biosensors: a review. Electroanal 22:1027–1036
13. Pumera M, Ambrosi A, Bonanni A, Chng ELK, Poh HL (2010) Graphene for electrochemical sensing and biosensing. TrAC, Trends Anal Chem 29:954–965
14. Wang K, Liu Q, Guan QM, Wu J, Li HN, Yan JJ (2011) Enhanced direct electrochemistry of glucose oxidase and biosensing for glucose via synergy effect of graphene and CdS nanocrystals. Biosens Bioelectron 26:2252–2257
15. Peng J, Gao W, Gupta BK, Liu Z, Romero-Aburto R, Ge LH, Song L, Alemany LB, Zhan XB, Gao GH, Vithayathil SA, Kaipparettu BA, Marti AA, Hayashi T, Zhu JJ, Ajayan PM (2012) Graphene quantum dots derived from carbon fibers. Nano Lett 12:844–849
16. Zhao XM, Zhou SW, Jiang LP, Hou WH, Shen QM, Zhu JJ (2012) Graphene–CdS nanocomposites: facile one-step synthesis and enhanced photoelectrochemical cytosensing. Chem Eur J 18:4974–4981
17. Shen QM, Zhou SW, Zhao XM, Jiang LP, Hou WH, Zhu JJ (2012) Anatase TiO2 nanoparticle–graphene nanocomposites: one-step preparation and their enhanced direct electrochemistry of hemoglobin. Anal Methods 4:619–622
18. Service RF (1998) Coming soon: the pocket DNA sequencer. Science 282:399–401
19. Staudt LM (2001) Gene expression physiology and pathophysiology of the immune system. Trends Immunol 22:35–40
20. Drummond TG, Hill MG, Barton JK (2003) Electrochemical DNA sensors. Nat Biotechnol 21:1192–1199
21. Ji HX, Yan F, Lei JP, Ju XH (2012) Ultrasensitive electrochemical detection of nucleic acids by template enhanced hybridization followed with rolling circle amplification. Anal Chem 84:7166–7171
22. Xu Q, Wang JH, Wang Z, Yin ZH, Yang Q, Zhao YD (2008) Interaction of CdTe quantum dots with DNA. Electrochem Commun 10:1337–1339
23. Yin CX, Yang T, Zhang W, Zhou XD, Jiao K (2010) Electrochemical biosensing for dsDNA damage induced by PbSe quantum dots under UV irradiation. Chinese Chem Lett 21:716–719

24. Huang HP, Li JJ, Tan YL, Zhou JJ, Zhu JJ (2010) Quantum dot-based DNA hybridization by electrochemiluminescence and anodic stripping voltammetry. Analyst 135:1773–1778
25. Chen JH, Zhang J, Yang HH, Fu FF, Chen GN (2010) A strategy for development of electrochemical DNA biosensor based on site-specific DNA cleavage of restriction endonuclease. Biosens Bioelectron 26:144–148
26. Wang J, Liu GD, Merkoi A (2003) Electrochemical coding technology for simultaneous detection of multiple DNA targets. J Am Chem Soc 125:3214–3215
27. Ellington AD, Szostak JW (1990) In vitro selection of RNA molecules that bind specific ligands. Nature 346:818–822
28. Tuerk C, Gold L (1990) Systematic evolution of ligands by exponential enrichment: RNA ligands to bacteriophage T4 DNA polymerase. Science 249:505–510
29. Gold L, Polisky B, Uhlenbeck O, Yarus M (1995) Diversity of oligonucleotide functions. Annu Rev Biochem 64:763–797
30. Hesselberth J, Robertson MP, Jhaveri S, Ellington AD (2000) In vitro selection of nucleic acids for diagnostic applications. Rev Mol Biotechnol 74:15–25
31. Smith JE, Medley CD, Tang ZW, Shang DH, Lofton C, Tan WH (2007) Aptamer-conjugated nanoparticles for the collection and detection of multiple cancer cells. Anal Chem 79:3075–3082
32. Zhang SS, Xia JP, Li XM (2008) Electrochemical biosensor for detection of adenosine based on structure-switching aptamer and amplification with reporter probe DNA modified Au nanoparticles. Anal Chem 80:8382–8388
33. Lu Y, Li XC, Zhang LM, Yu P, Su L, Mao LQ (2008) Aptamer-based electrochemical sensors with aptamer-complementary DNA oligonucleotides as probe. Anal Chem 80:1883–1890
34. Jayasena SD (1999) Aptamers: an emerging class of molecules that rival antibodies in diagnostics. Clin Chem 45:1628–1650
35. Breaker RR (1997) DNA aptamers and DNA enzymes. Curr Opin Chem Biol 1:26–31
36. Chen Y, Wang MS, Mao CD (2004) An autonomous DNA nanomotor powered by a DNA enzyme. Angew Chem Int Ed 43:3554–3557
37. Yoshizumi J, Kumamoto S, Nakamura M, Yamana K (2008) Target-induced strand release (TISR) from aptamer–DNA duplex: a general strategy for electronic detection of biomolecules ranging from a small molecule to a large protein. Analyst 133:323–325
38. Wang J, Wang L, Liu X, Liang Z, Song S, Li W, Li G, Fan CH (2007) A gold nanoparticle-based aptamer target binding readout for ATP assay. Adv Mater 19:3943–3946
39. Zhao Q, Li XF, Le XC (2008) Aptamer-modified monolithic capillary chromatography for protein separation and detection. Anal Chem 80:3915–3920
40. Tombelli S, Minunni M, Luzi E, Mascini M (2005) Aptamer-based biosensors for the detection of HIV-1 Tat protein. Bioelectrochem 67:135–141
41. Zhou JJ, Huang HP, Xuan J, Zhang JR, Zhu JJ (2010) Quantum dots electrochemical aptasensor based on three-dimensionally ordered macroporous gold film for the detection of ATP. Biosens Bioelectron 26:834–840
42. Dong XY, Mi XN, Zhao WW, Xu JJ, Chen HY (2011) CdS nanoparticles functionalized colloidal carbon particles: Preparation, characterization and application for electrochemical detection of thrombin. Biosens Bioelectron 26:3654–3659
43. Zhang HX, Jiang BY, Xiang Y, Zhang YY, Chai YQ, Yuan R (2011) Aptamer/quantum dot-based simultaneous electrochemical detection of multiple small molecules. Analy Chim Acta 688:99–103
44. Hansen JA, Wang J, Kawde A, Xiang Y, Gothelf KV, Collins G (2006) Quantum-dot/aptamer-based ultrasensitive multi-analyte electrochemical biosensor. J Am Chem Soc 128:2228–2229
45. Li YJ, Ma MJ, Zhu J-J (2012) Dual-signal amplification strategy for ultrasensitive photoelectrochemical immunosensing of α-fetoprotein. Anal Chem 84:10492–10499
46. Wang GL, Xu JJ, Chen HY, Fu SZ (2009) Label-free photoelectrochemical immunoassay for α-fetoprotein detection based on TiO2/CdS hybrid. Biosens Bioelectron 25:791–796
47. Wang GL, Yu PP, Xu JJ, Chen HY (2009) A label-free photoelectrochemical immunosensor based on water-soluble CdS quantum dots. J Phys Chem C 113:11142–11148

48. Zhang XR, Zhao YQ, Zhou HR, Qu B (2011) A new strategy for photoelectrochemical DNA biosensor using chemiluminescence reaction as light source. Biosens Bioelectron 26:2737–2741
49. Baş D, Boyacı İH (2011) Photoelectrochemical competitive DNA hybridization assay using semiconductor quantum dot conjugated oligonucleotides. Anal Bioanal Chem 400:703–707
50. Willner I, Patolsky F, Wasserman J (2001) Photoelectrochemistry with controlled DNA-cross-linked CdS nanoparticle arrays. Angew Chem Int Ed 40(10):1861–1864
51. Porath D, Cuniberti G, Felice RD (2004) Charge transport in DNA-based devices. Top Curr Chem 37:183–227
52. De Pablo PJ, Moreno-Herrero F, Colchero J (2000) Absence of dc-conductivity in λ-DNA. Phys Rev Lett 85:4992–4995
53. OKNeill MA, Barton JK (2004) DNA charge transport: conformationally gated hopping through stacked domains. J Am Chem Soc 126:11471–11483
54. Drummond TG, Hill MG, Barton JK (2003) Electrochemical DNA sensors. Nat Biotechnol 21:1192–1199
55. Gill R, Patolsky F, Katz E, Willner I (2005) Electrochemical control of the photocurrent direction in intercalated DNA/CdS nanoparticle systems. Angew Chem Int Ed 44:4554–4557
56. Freeman R, Gill R, Beissenhirtz M, Willner I (2007) Self-assembly of semiconductor quantum-dots on electrodes for photoelectrochemical biosensing. Photochem Photobiol Sci 6:416–422
57. Zhao WW, Wang J, Xu JJ, Chen HY (2011) Energy transfer between CdS quantum dots and Au nanoparticles in photoelectrochemical detection. Chem Commun 47:10990–10992
58. Zhao WW, Yu PP, Shan Y, Wang J, Xu JJ, Chen HY (2012) Exciton-plasmon interactions between CdS quantum dots and Ag nanoparticles in photoelectrochemical system and its bio-sensing application. Anal Chem 84:5892–5897
59. Golub E, Niazov A, Freeman R, Zatsepin M, Willner I (2012) Photoelectrochemical bio-sensors without external irradiation: probing enzyme activities and DNA sensing using Hemin/GQuadruplex-stimulated chemiluminescence resonance energy transfer (CRET) generation of photocurrents. J Phys Chem C 116:13827–13834